Reactive Oxygen Species and the Cardiovascular System

Colloquium Series on Integrated Systems Physiology: From Molecule to Function to Disease

Editors

D. Neil Granger, *Louisiana State University Health Sciences Center*

Joey P. Granger, *University of Mississippi Medical Center*

Physiology is a scientific discipline devoted to understanding the functions of the body. It addresses function at multiple levels, including molecular, cellular, organ, and system. An appreciation of the processes that occur at each level is necessary to understand function in health and the dysfunction associated with disease. Homeostasis and integration are fundamental principles of physiology that account for the relative constancy of organ processes and bodily function even in the face of substantial environmental changes. This constancy results from integrative, cooperative interactions of chemical and electrical signaling processes within and between cells, organs and systems. This eBook series on the broad field of physiology covers the major organ systems from an integrative perspective that addresses the molecular and cellular processes that contribute to homeostasis. Material on pathophysiology is also included throughout the eBooks. The state-of the-art treatises were produced by leading experts in the field of physiology. Each eBook includes stand-alone information and is intended to be of value to students, scientists, and clinicians in the biomedical sciences. Since physiological concepts are an ever-changing work-in-progress, each contributor will have the opportunity to make periodic updates of the covered material.

Published titles

(for future titles please see the website, www.morganclaypool.com/page/lifesci)

Reactive Oxygen Species and the Cardiovascular System
Augusto C. Montezano and Rhian M. Touyz
www.morganclaypool.com

ISBN: 9781615043620 paperback

ISBN: 9781615043637 ebook

DOI: 10.4199/C00043ED1V01Y201112ISP032

A Publication in the

COLLOQUIUM SERIES ON INTEGRATED SYSTEMS PHYSIOLOGY: FROM MOLECULE TO FUNCTION TO DISEASE

Lecture #32

Series Editors: D. Neil Granger, LSU Health Sciences Center, and Joey P. Granger, University of Mississippi Medical Center

Series ISSN

ISSN 2154-560X   print

ISSN 2154-5626   electronic

Reactive Oxygen Species and the Cardiovascular System

Augusto C. Montezano and Rhian M. Touyz
University of Ottawa and University of Glasgow

COLLOQUIUM SERIES ON INTEGRATED SYSTEMS PHYSIOLOGY:
FROM MOLECULE TO FUNCTION TO DISEASE #32

ABSTRACT

Reactive oxygen species (ROS) influence various physiological processes including host defense, hormone biosynthesis, and cellular signaling. Increased ROS production (oxidative stress) is implicated in many diseases of the cardiovascular system, including hypertension, atherosclerosis, cardiac failure, stroke, diabetes, and kidney disease. ROS are produced throughout the cardiovascular system, in the kidney and central and peripheral nervous system. A major source for cardiovascular, renal, and neural ROS is a family of non-phagocytic NAD(P)H oxidases, including the prototypic Nox2 homologue-based NAD(P)H oxidase, as well as other NAD(P)H oxidases, such as Nox1 and Nox4. Other possible sources include mitochondrial electron transport enzymes, xanthine oxidase, cyclooxygenase, lipoxygenase, and uncoupled nitric oxide synthase (NOS). NAD(P)H oxidase-derived ROS is important in regulating endothelial function and vascular tone and oxidative stress is implicated in endothelial dysfunction, inflammation, hypertrophy, apoptosis, migration, fibrosis, angiogenesis and rarefaction, important processes involved in vascular remodeling in cardiovascular disease. These findings have evoked considerable interest because of the possibilities that therapies targeted against non-phagocytic NAD(P)H oxidase to decrease ROS generation and/or strategies to increase nitric oxide (NO) availability and antioxidants may be useful in minimizing vascular injury and thereby prevent or regress target organ damage associated with hypertension and other cardiovascular diseases.

KEYWORDS

superoxide, hydrogen peroxide, Nox, heart, vascular system, kidneys, blood pressure, vascular remodeling, inflammation, antioxidant.

Contents

CHAPTER 1

Introduction

Oxygen is essential for life, while at the same time may cause cell harm. This "oxygen paradox" relates to the fact that atomic oxygen is an unstable free radical while molecular oxygen is a stable biradical [1]. The univalent reduction of O_2 leads to generation of reactive species, known as reactive oxygen species (ROS), including superoxide radical anion ($\bullet O_2^-$), hydroxyl radical (HO$\bullet$) and peroxyl radicals (ROO$\bullet$) and to nonradical derivatives, such as hydrogen peroxide (H_2O_2), peroxynitrite (ONNO$^-$) and hypochlorous acid (HOCL). ROS were originally considered to induce negative and injurious cellular effects through oxidative damage of DNA, proteins and lipids, causing apoptosis, necrosis and cell death. However, growing evidence indicates that ROS also have important positive and physiologically significant actions such as the induction of host defense genes, regulation of signaling molecules, activation of transcription factors and stimulation of ion transport channels [2–5]. ROS play an important role in regulating the cardiovascular, renal and nervous systems, all critically involved in blood pressure regulation. In the cardiovascular system, ROS influence cardiac and endothelial function and modulate vascular tone. When in excess, ROS play a pathophysiological role in endothelial dysfunction, inflammation, hypertrophy, proliferation, apoptosis, migration, fibrosis, angiogenesis and rarefaction, important processes underlying cardiovascular remodeling in hypertension and other cardiovascular diseases [6–8].

Subcellular processes whereby ROS induce cardiovascular and renal oxidative damage involve activation of redox-sensitive signaling pathways [9–11]. Superoxide anion and H_2O_2, which are increasingly being recognized as signaling molecules, stimulate mitogen-activated protein (MAP) kinases, tyrosine kinases, ion channels and transcription factors (NFκB, AP-1 and HIF-1) and inactivate protein tyrosine phosphatases [12–14]. ROS also increase intracellular-free Ca^{2+} concentration ($[Ca^{2+}]_i$) and upregulate protooncogene, profibrotic and proinflammatory gene expression and activity [15–17]. These phenomena occur through multiple processes, including oxidative modification of proteins, alteration of key amino acid residues, induction of protein dimerization and interaction with metal complexes such as Fe–S moieties [18, 19]. Changes in the intracellular redox state through glutathione and thioredoxin systems may also influence intracellular signaling events [20, 21].

Although ROS/oxidative stress has been implicated in many pathologies, the focus here will be on hypertension. Most experimental models of hypertension exhibit oxidative stress (increased bioavailability of ROS), including genetic forms (SHR, SHRSP), surgically induced (2K1C, aortic banding), endocrine-induced (Ang II, aldosterone, DOCA) and diet-induced hypertension (salt, fat) [22–28]. Mice deficient in ROS-generating enzymes have lower blood pressure compared with wild-type counterparts, and Ang II infusion fails to induce hypertension in these mice [29, 30]. Since inhibition of ROS-generating enzymes, anti-oxidants and ROS-scavengers reduce blood pressure whereas pro-oxidants increase blood pressure, it has been suggested that ROS are causally associated with hypertension, at least in animal models.

In human hypertension, biomarkers of systemic oxidative stress, including levels of plasma thiobarbituric acid-reactive substances (TBARS) and 8-epi-isoprostanes, are increased [31–33]. Factors contributing to oxidative stress in human hypertension include decreased antioxidant activity, reduced levels of ROS scavengers and activation of ROS-generating enzymes [34–36]. A causal link between ROS and high blood pressure has not yet been definitively demonstrated in humans. Only a few small clinical studies showed a blood pressure-lowering effect of anti-oxidants [37–39], whereas many large anti-oxidant clinical trials failed to demonstrate any cardiovascular benefit and blood pressure reduction [40–42]. However, it should be stressed that these large trials did not examine blood pressure as an endpoint. However, what is evident is that oxidative stress plays a critical role in the molecular and cellular processes underlying cardiovascular and renal damage in hypertension and that high blood pressure itself can contribute to oxidative stress. A greater understanding of the (patho)biology of ROS may lead to new insights and novel diagnostics and treatments for hypertension and other cardiovascular diseases.

. . . .

CHAPTER 2

Redox Molecules

The dynamics and physical properties of ROS dictate the function they will play within the body. The more reactive the species, the shorter is the half life and the more rapidly it will interact with other molecules. Moreover, the distance that the reactive species travels will be much shorter due to its highly reactive nature. The reverse is true for less reactive species, such as H_2O_2. Hydrogen peroxide, which is not an extremely reactive ROS, can even cross cellular membranes and exert its actions outside of the site of its production. ROS can also be electrically charged or electrically neutral, hydrophobic or hydrophilic. Such characteristics can dictate their ability to cross membranes and/or to move in the environment between aqueous and lipophilic environments.

Reactive oxygen species are produced as intermediates in reduction–oxidation (redox) reactions leading from O_2 to H_2O (Figure 1) [43, 44]. The major mechanism for ROS generation begins with reduction of O_2 by the addition of one electron, to generate $•O_2^-$, considered the primary ROS. Superoxide anion interacts with other molecules to produce secondary ROS, directly or through enzyme- or metal-catalyzed reactions [3, 45, 46]. Reduction of $•O_2^-$ leads to formation of H_2O_2, which is further converted to secondary metabolites such as highly reactive HO• [46]. Although the favored reaction is generation of H_2O_2, $•O_2^-$ also reacts with nitric oxide (NO•) to form ONOO–, with transition metals, such as iron found in iron/sulfur center-containing proteins or it may be protonated to the hydroperoxyl radical ($HO_2•$). $HO_2•$ is particularly important in lipid peroxidation and atherogenesis.

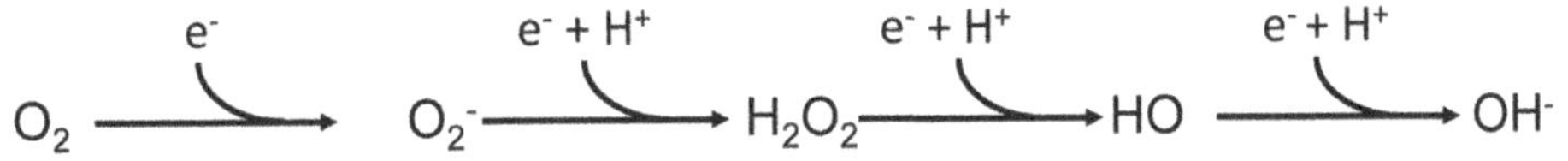

FIGURE 1: Generation of superoxide anion and secondary reactive oxygen species from molecular oxygen.

2.1 SUPEROXIDE AND HYDROGEN PEROXIDE

Of the ROS generated in cardiovascular cells, $\bullet O_2^-$ and H_2O_2 appear to be especially important. In cells, $\bullet O_2^-$ originates from mitochondria, where $\bullet O_2^-$ is produced by electrons leaking between complexes 1 and III of the electron-transport chain [47]. Superoxide anion is also generated as a product of enzymatic activation where enzymes use NADPH to reduce O_2. The prototype enzyme is NADPH oxidase (Nox), which is a professional ROS-generating enzyme, since its primary function is to produce $\bullet O_2^-$ in a highly regulated manner [48]. Many other proteins also generate ROS; however, this is usually as a byproduct of normal function.

In biological systems, $\bullet O_2^-$ is short-lived owing to its rapid dismutation to H_2O_2. Dismutation can be spontaneous (rate constant = 8×10^4/mol/sec) or enzymatic via superoxide dismutase (SOD) (rate constant = 2×10^9/mol/sec) [43, 49]. Superoxide reacts with NO$\bullet$ to form ONOO- with a rate constant of $4–16 \times 10^9$/mol/sec [50] resulting in NO$\bullet$ quenching and resultant decreased NO$\bullet$ bioavailability. In addition $\bullet O_2^-$ reacts with FeS_4 or with protein thiols such as cysteine residues [51, 52]. Although $\bullet O_2^-$ has the capacity to react with many molecules, the preferred reaction is dismutation to H_2O_2 because of the fast reaction rate of SOD. Three mammalian SOD isoforms have been identified: copper/zinc SOD (SOD1), mitochondrial SOD (Mn SOD, SOD2) and extracellular SOD (EC-SOD, SOD3) [53–56]. The major vascular SOD is EC-SOD (56).

The negative charge on $\bullet O_2^-$ makes it unable to cross cellular membranes except possibly through ion channels, such as chloride channel 3 (CLC-3). CLC-3 transports $\bullet O_2^-$ out of endosomes into the cytoplasm in endothelial cells [57, 58] and has been implicated to play an important role in vascular smooth muscle cell regulation [59]. Hydrogen peroxide has a longer lifespan than $\bullet O_2^-$, is relatively stable and is easily diffusible within and between cells. The main source of H_2O_2 in vascular tissue is the dismutation of $\bullet O_2^-$: $2\bullet O_2^- + 2H+ \rightarrow H_2O_2 + O_2$. H2O2 is tightly regulated by intracellular and extracellular enzymes, including catalase, glutathione peroxidases, thioredoxin and other peroxredoxins, which convert H_2O_2 to water and O_2 and other metabolites. Although both $\bullet O_2^-$ and H_2O_2 have been suggested to act as signaling molecules, it is mainly H_2O_2 that is considered a signaling molecule because of its relative stability, tight regulation, subcellular localization and ability to react reversibly with cysteine residues [60, 61]. The ROS-specific effects are mediated in large part through the oxidative modification of specific cysteine residues within redox-sensitive target proteins [61, 62]. ROS regulate many physiological functions and when dysregulated, ROS signaling contributes to pathological conditions.

The distinct chemical properties between $\bullet O_2^-$ and H_2O_2 and their different sites of distribution mean that different species of ROS activate diverse signaling pathways, which lead to divergent, and potentially opposing, biological responses. For example, in the vasculature, increased $\bullet O_2^-$ levels inactivate the vasodilator NO leading to endothelial dysfunction and vasoconstriction [63, 64], whereas H_2O_2 acts as a direct vasodilator in some vascular beds, including cerebral, coronary and mesenteric arteries [65–67].

2.2 REACTIVE NITROGEN SPECIES (RNS)

The main RNS is NO, which is a potent vasodilator and responsible for endothelium-dependent vasorelaxation [68, 69]. Nitric oxide is also anti-inflammatory and anti-mitogenic and, because of its unique properties, was named the "molecule of the year" in 1992 [70]. Nitric oxide is produced by nitric oxide synthase (NOS), of which there are 3 isoforms: endothelial (eNOS), neuronal (nNOS) and inducible (iNOS) [71]. Nitric oxide can also be generated by other redox enzymes, including xanthine oxidase [72]. Metabolism and reactivity of NO• leads to the generation of other RNS, including ONOO– and nitrogen dioxide (•NO_2) [73, 74]. Peroxynitrite and nitroxidative stress are implicated in various aspects of nitrooxidative cellular damage since peroxynitrite also yields secondary one electron oxidants [73, 74].

Reduced NO bioavailability, increased NOS-derived ROS and ONOO– formation are critically involved in vascular, cardiac and renal dysfunction and inflammation associated with cardiovascular disease. Hence, restoring and conserving adequate NO signaling in the cardiovascular system by modulating eNOS through increasing the bioavailability of its substrate and cofactors and enhancing its transcription has been suggested as a promising therapeutic intervention [75].

· · · ·

CHAPTER 3

Production and Metabolism of ROS in the Cardiovascular System

ROS are generated by all vascular cell types, including endothelial, smooth muscle and adventitial cells (fibroblasts, adipocytes), and can be formed by many enzymes. In the heart, cardiomyocytes and fibroblasts produce ROS. Enzymatic sources of ROS important in cardiovascular disease and hypertension are xanthine oxidoreductase, uncoupled NO synthase (NOS), mitochondrial respiratory enzymes, and nicotinamide adenine dinucleotide phosphate (NADPH) oxidase (Figure 2) [76–79].

3.1   XANTHINE OXIDASE

Xanthine oxidase (XO) and xanthine dehydrogenase (XDH) are interconvertible forms of the same enzyme, known as xanthine oxidoreductase. Physiologically, XO and XDH participate in many biochemical reactions, with the primary role being degradation of purines and the conversion of hypoxanthine to xanthine and xanthine to uric acid. As a byproduct in the purine degradation pathway, XO oxidizes NADH to form $\bullet O_2^-$ and H_2O_2. In the vascular wall, XO-derived $\bullet O_2^-$ reacts rapidly with NO to form ONOO-, which can lead to a negative feedback of the enzyme [76, 80, 81]. Uric acid, which has antioxidant potential, also acts as a feedback inhibitor of XO. Xanthine oxidase is expressed in vascular cells, it circulates in the plasma and it binds to endothelial cell extracellular matrix. Although xanthine oxidase-derived $\bullet O_2^-$ has been studied mainly in the context of cardiac disease and atherosclerosis, there is evidence suggesting involvement in hypertension. Spontaneously hypertensive rats (SHR) and DOCA-salt hypertensive rats demonstrate elevated levels of endothelial XO and increased ROS production, which are associated with increased arteriolar tone [82]. This may be mediated, in part, through an adrenal pathway, because adrenalectomy reduces XO expression [83]. Endothelial dysfunction in transgenic rats with overexpression of renin and angiotensinogen has also been associated with increased XO activity [84]. In addition to effects on the vasculature, XO may play a role in end-organ damage in hypertension. In experimental models of hypertension, XO activity is increased in the kidney. Long-term inhibition of XO with allopurinol in SHR reduced renal XO activity without lowering blood pressure, indicating that the

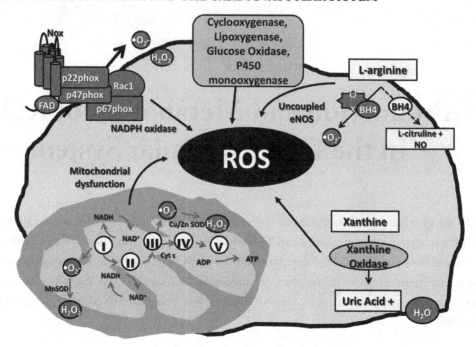

FIGURE 2: Sources of cellular ROS in the cardiovascular system. Production of different ROS is regulated by complex control systems The enzymes that produce ROS in the cardiovascular system are diverse and include: 1) xanthine oxidase, uncoupled eNOS, 3) mitochondrial oxidases, 4) NADPH oxidases. Xanthine oxidase catalyses the oxidation of xanthine to uric acid, by reducing oxygen to both superoxide and hydrogen peroxide, Uncoupled eNOS, produces NO, which can interact with superoxide to form peroxynitrite. In the absence of L-arginine or tetrahydrobiopterin, the electrons in the enzyme reduce molecular oxygen to superoxide, instead of NO. Mitochondrial production of ROS occurs when 95% of molecular oxygen is reduced by 4 electrons to yield two molecules of water, via the electron transport chain complexes (I–IV). In aerobic conditions, 1–2% of superoxide produced is spilled and rapidly neutralized by antioxidant enzymes. Damaged or dysfunctional mitochondria overgenerate superoxide, creating a state of redox imbalance and consequent oxidative stress. NADPH oxidase (Nox) comprises a family of 7 enzymes, where the generation of ROS is the main and only function. For these enzymes to be biologically active, they require a number of subunits to assemble at the membrane in order to form ROS. Other enzymes that generate ROS are cyclooxygenase, lipoxygenase, glucose oxidase and P450 mono-oxidases.

increased renal ROS production was a consequence of hypertension rather than a contributing factor [85]. The finding that allopurinol can improve cardiac and renal hypertrophy in SHR and slow the progression of renal disease in patients with chronic kidney disease and hypertension [86], while having a minimal impact on blood pressure [87] supports a role for XO in hypertensive end-organ damage. This may be mediated through direct vascular effects of XO-produced uric acid [88]. To further support a role for XO in the pathogenesis of hypertension, allopurinol decreased blood pressure in hyperuricemic adolescents with newly diagnosed hypertension [89]. However, it still remains unclear whether $\bullet O_2^-$ or uric acid is the primary factor involved in XO-sensitive hypertension.

3.2   UNCOUPLED NITRIC OXIDE SYNTHASE

Under physiological conditions, nitric oxide synthase (NOS), in the presence of cofactors L-arginine and tetrahydrobiopterin (BH4), produces NO. In the absence of these cofactors, because of oxidative destruction or downregulation of GTP cyclohydrolase-1, which is the rate-limiting enzyme in BH4 production, uncoupled NOS produces $\bullet O_2^-$ rather than NO [90, 91]. All three NOS isoforms are capable of 'uncoupling' that leads to the preferential formation of $\bullet O_2^-$ [90, 91]. eNOS uncoupling has been demonstrated in DOCA-salt-induced hypertension and in SHR [92, 93] and has been implicated in atherosclerosis and endothelial dysfunction in low-density lipoprotein receptor-deficient mice (LDLR-/-) fed a high salt, high-fat diet [94, 95]. Dysfunctional eNOS is also important in cardiac remodeling from pressure overload. In mice subjected to proximal aortic constriction, oral BH4 prevented progressive chamber dilation and dysfunction, reversed fibrosis and hypertrophy, and improved myocyte function and calcium handling [96]. This was associated with eNOS recoupling and reduced oxidative stress. Whether effects of uncoupled NOS are due to increased $\bullet O_2^-$ generation or to decreased NO bioavailability still remain unclear [97]. Nevertheless, BH4 has been suggested as a treatment modality for hypertension, endothelial dysfunction, atherosclerosis, diabetes, cardiac hypertrophic remodeling and heart failure [98–100]. While previously difficult to use clinically because of chemical instability and cost, newer methods to synthesize stable BH4 suggest its novel potential as a therapeutic agent [101]. In fact, some classical antihypertensive drugs, including the beta blocker nebivolol, have been shown to induce effects by preventing eNOS uncoupling [102].

3.3   MITOCHONDRIAL RESPIRATORY ENZYMES

Mitochondria are energy-producing organelles with multiple functions including the regulation of cytosolic Ca^{2+} levels and tissue O_2 gradients, H_2O_2 signaling, modulation of apoptosis and integration of cellular responses to numerous stimuli [103–108]. Mitochondria are considered as the main source of cellular ROS, with more than 95% of all O_2 consumed by cells being reduced by four electrons to yield two molecules of H_2O via the mitochondrial electron transport [109–111].

More than 95% of O_2 consumed by cells is reduced by four electrons to yield two molecules of H_2O via mitochondrial electron transport chain complexes (I–IV), with 1–2% of the electron flow leaking onto O_2 to form $\bullet O_2^-$ under normoxic conditions [112]. Mitochondrial ROS production is modulated by many factors including mitochondrial electron transport chain efficiency [113], mitochondrial antioxidant content [114], local oxygen and NO concentrations [115, 116], availability of metabolic electron donors [117], uncoupling protein (UCP) activity [118], cytokines and vasoactive agonists [119–122].

Ang II and ET-1 stimulate mitochondrial ROS generation in endothelial and vascular smooth muscle cells and in rat aorta *in vivo* [123–132]. Mechanisms whereby these vasoactive agents stimulate mitochondrial ROS production are unclear but could involve opening of mitochondrial potassium channels (mitoK$_{ATP}$) [133] and mitochondrial permeability transition (MPT) [134, 135]. Interestingly, Ang II may interact directly with mitochondria as evidenced by studies demonstrating that labeled ^{125}I–Ang II is detectable in cardiac, brain and smooth muscle mitochondria [119, 136].

Alterations in mitochondrial biogenesis are associated with mitochondrial dysfunction and mitochondrial oxidative stress. Impaired activity and/or decreased expression of mitochondrial electron transport chain complexes I, III and IV have been implicated in vascular aging, cardiovascular disease, diabetes and other chronic diseases [137–142]. An association between mitochondrial dysfunction and blood pressure has been reported in human and experimental hypertension [142–145]. Ang II-sensitive hypertension is also linked to mitochondrial-derived oxidative stress since AT$_1$ receptor blockade attenuates H_2O_2 production [146] and mitochondrial dysfunction in SHR, and in mice, Ang II infusion is associated with decreased expression of cardiac mitochondrial electron transport genes. In DOCA-salt hypertension, mitochondrial-derived ROS plays an important role in oxidative vascular damage, an effect mediated via ET-1/ETA receptors [147–149]. Chan and coworkers [150] have provided new evidence that mitochondrial dysfunction and mitochondrial-localized ROS production in the central nervous system is important in cardiovascular function. They demonstrated a relationship between decreased activity of complex I and complex III and increased ROS production. When electron transport was re-established, ROS formation was decreased, and blood pressure was reduced [151]. Clinically, Yang et al. showed that mitochondrial heritability for systolic blood pressure was about 5% and mitochondrial effects may account for 35% of hypertensive pedigrees [152]. In African Americans with hypertension-associated end-stage renal disease, mitochondrial-DNA mutations in the kidneys have been identified [153].

3.4   NOX-FAMILY NAD(P)H OXIDASES

NADPH oxidase was originally considered to be expressed only in phagocytic cells involved in host defense and innate immunity. It is now evident that there is a family of NAD(P)H oxidases, based on homologues of the catalytic subunit gp91phox that are functionally active in non-phagocytic cells. The new homologues, along with gp91phox are the designated Nox family of NAD(P)H oxidases [154] and are important in vascular ROS production (Figure 3). The prototypical gp91phox-containing phagocytic NAD(P)H oxidase (now termed Nox2) comprises 5 subunits: p47phox ("phox" stands for *ph*agocyte *ox*idase), p67phox, p40phox, p22phox, and the catalytic

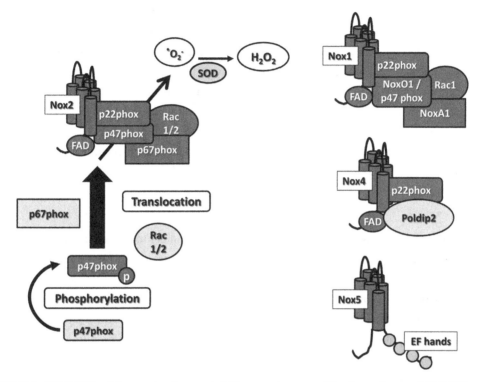

FIGURE 3: NADPH oxidase activation and differences between Nox homologues. NADPH oxidase comprises a complex of membrane and cytosolic subunits. Nox2 is the classical prototype. Membrane proteins are p22phox and the Nox subunit, and form a non-covalent heterodimer. These proteins possess the electron transport apparatus and may act as a physical conduit for the electron transfer that occurs across the membrane. The cytosolic proteins (p47phox, p67phox, NoxO1, NoxA1, rac 1/2) are cofactors for enzymatic activity and are used to initiate and/or regulate electron transfer. In order to Nox2 be activated, p47phox is phosphorylated and translocates from the cytosol to the membrane with the other cytosolic subunits (p67phox and rac1/2). Nox1 can also be activated in a similar way to Nox2, but possess other cytosolic subunits, such as NoxO1 and NoxA1. Nox4 does not require any cytosolic subunit to be activated. Nox4 is constitutively active in cells and its activity is controlled by Poldip2. On the other hand, Nox5 activation is not dependent on any subunit, but because of calcium-binding domains (EF hands), its activity is controlled by calcium and calmodulin.

subunit gp91phox [155]. In basal conditions p47phox, p67phox and p40phox, exist in the cytosol, whereas p22phox and gp91phox are in the membrane, where they occur as a heterodimeric flavo-protein, (cytochrome b558). Upon stimulation, p47phox and p67phox form a complex that trans-locates to the membrane, where it associates with cytochrome b558 to assemble the active oxidase, which transfers electrons from the substrate to O_2 forming $\bullet O_2^-$ [156]. Activation also requires Rac 2 (or Rac 1) and Rap 1A.

The mammalian Nox family comprises seven members: Nox1, Nox2, Nox3, Nox4, Nox5, Duox1 and Duox2 [154–157]. All are transmembrane proteins that have a core catalytic subunit (Nox) and numerous regulatory subunits. Nox1, Nox2, Nox4 and Nox5 have been identified in car-diovascular and renal tissue [158]. Hyperactivation of Noxs leads to excessive ROS generation that disrupts redox networks, normally regulated by thiol-dependent antioxidant systems. This results in oxidative stress, triggering molecular processes, which in the vasculature, contributes to vascular in-jury. The Noxs are functionally active in many tissue systems that contribute to blood pressure regu-lation and development of hypertension (Figure 4). Noxs have been extensively reviewed [159–162] and only an overview of recent developments is discussed here.

3.4.1   Nox1

Nox1 is expressed in vascular and cardiac cells [163]. It localizes to the cell membrane, caveolae/lipid rafts and endosomes. Nox1 is similar to Nox2 in that it requires p22phox, p47phox (or its homo-logue NoxO1 (Nox organizer 1)) p67phox (or its homologue NoxA1 (Nox activator 1)) and Rac1 for its activity. New Nox1 regulators have been identified, Tks4 and Tks5, which resemble p47phox and NoxO1 and which interact with NoxA1 [164]. Nox1 localizes with p22phox and is expressed at low levels in physiological conditions. Nox1-derived $\bullet O_2^-$ is increased in a stimulus-dependent manner, involving complex interactions between regulatory subunits and the redox chaperone pro-tein disulfide isomerase (PDI) [165, 166]. Nox1 has been implicated in vascular smooth muscle cell migration, proliferation and extracellular matrix production, effects mediated by cofilin [167].

In cultured endothelial and vascular smooth muscle cells Nox1 is upregulated by mechani-cal factors (shear stress), vasoactive agents (Ang II, aldosterone), growth factors (EGF, PDGF) [168, 169]. Ang II-induced induction of Nox1 may involve mitochondria, suggesting an interaction between Nox1 and mitochondria, possibly through a Ca^{2+}-dependent mechanism [170]. Nox1 expression/activity is increased in the vasculature in models of cardiovascular disease including hypertension, atherosclerosis, diabetes and hypercholesterolemia [171]. Studies from Nox1 knock-out and transgenic mice suggest a possible role for Nox1 in acute, but not chronic, forms of Ang II-dependent hypertension [172, 173], in atherosclerosis, restenosis post injury, endothelial dys-function and stroke. Mice genetically deficient in Nox1 also display decreased expression of aortic AT_1R [174], which may contribute to blunted hypertensive effects of Ang II infusion in these

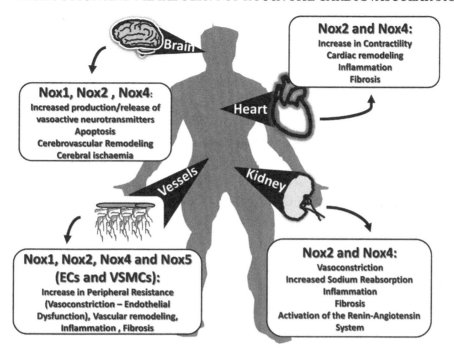

FIGURE 4: Tissue distribution of Nox homologues in the cardiovascular system. Different Nox homologues are expressed in the brain, heart, vessels and kidneys. The brain seems to express Nox1, Nox2 and Nox4; where activation of these enzymes is important to the control of neurotransmitter production/release, cell survival and cerebrovascular homeostasis. In the heart and kidneys, Nox2, and Nox4 have been characterized. They seem to play a role in cardiac contractility and remodeling, in the heart; and sodium balance control and Ang II production, in the kidney. Vessels express Nox1, Nox2, Nox4 and Nox5. ROS generation by NADPH oxidases is important in the control of vascular tone and remodeling.

mice. Although there is extensive experimental data implicating Nox1 in cardiovascular disease, there is little information in humans, although expression of Nox1 and NoxA1 is increased in human atherosclerotic vessels [175].

3.4.2   Nox2

Nox2 is the catalytic subunit of the respiratory burst oxidase in phagocytes, but is also expressed in cardiovascular cells [176]. Nox2 is unstable without p22phox and requires p47 phox, p67phox and Rac1/2 for its full activation. In neutrophils, Nox2 localizes to intracellular and plasma membranes, and in vascular cells, it also localizes with the cytoskeleton, lipid rafts/caveolae and in the perinuclear compartment. The Nox2 gene is inducible and is highly regulated by Ang II and stretch. Vascular Nox2, derived from resident macrophages or vascular cells, is upregulated in experimental hypertension [177, 178], atherosclerosis, ischemia–reperfusion injury and neointimal formation. Although Nox2 has been shown to be important in models of Ang II-infused hypertension, it does not seem to play a role in blood pressure elevation or cardiac hypertrophy in a model of chronic Ang II-dependent hypertension [180]. Nox2 is also implicated in stroke in experimental models. Nox2-deficient mice exhibit significant reduction in cerebral infarct size compared with wild-type controls (179). In humans, NADPH oxidase has been shown to play a role in endothelial function since patients with chronic granulomatous disease (CGD), who have an X-linked Nox2 mutation, exhibit a significant increase in forearm-mediated vasodilation with increased NO bioavailability [181], suggesting that Nox2-based NADPH oxidase influences endothelial function and NO biology in humans. In patients with CGD with mutations in Nox2 or p47phox endothelial ischemia/ reperfusion injury was blunted, indicating a role for NADH oxidase-derived ROS in human ischemia/reperfusion injury [182].

3.4.3   Nox4

Nox4, of which 4 splice variants have been identified (NOX4B, NOXC, NOX4D and NOX4E), is found in vascular cells, fibroblasts and osteoclasts and is abundantly expressed in the kidney [183, 184]. In vascular smooth muscle cells, Nox4 co-localizes with p22phox and vinculin in focal adhesions and has been implicated in cell migration, proliferation, tube formation, angiogenesis and cell differentiation [185]. Nox4 has been identified in the endoplasmic reticulum, mitochondria and nucleus of vascular cells. Nox 4 does not seem to require p47phox, p67phox, p40phox, or Rac for its activation, although Nox R1 and Poldip2, Nox4 binding proteins, have recently been shown to be important [186].

Unlike Nox1 and Nox2, Nox4 is constitutively active, producing primarily H_2O_2 rather than $\cdot O_2^-$ [187]. The difference in the species generated may underlie Nox-specific actions in cell signaling. Nox4 contributes to basal ROS production through its constitutive activity and to increased ROS generation when stimulated by Ang II, glucose, TNFα and growth factors. The pathological role of Nox4 is unclear, although it has been implicated in hypertension, atherosclerosis and cardiovascular and renal complications of diabetes and in remodeling of pulmonary arteries in pulmonary hypertension [188]. Nox4-derivd ROS has also been suggested in cellular senescence and aging [189] and in insulin-mediated differentiation of adipocytes [190]. Recent studies demonstrated that Nox4 may have protective effects. In mice with a genetic deletion of Nox4 or a cardiomyocyte-targeted overexpression of Nox4, basal cardiac function was normal in both models, but Nox4-null animals developed exaggerated contractile dysfunction, hypertrophy and cardiac dilatation during exposure to chronic overload whereas Nox4-transgenic mice were protected [191]. Nox4-derived H_2O_2 may act as a vasodilator in some vascular beds, which could explain why mice with targeted endothelial Nox4 overexpression in the endothelium exhibits lower blood pressure and improved endothelium-dependent vasodilation compared with wild type controls [192]. The exact (patho)physiological role of Nox4 in the cardiovascular system remains unclear because many of the *in vivo* studies interrogating Nox4 were performed in transgenic mice where Nox4 was up- or down-regulated.

3.4.4   Nox5

Nox5 is the most recently identified of the Nox enzymes and has unique features compared with other family members. Nox5 is a Ca^{2+}-sensitive homologue found in testes, spleen and lymphoid tissue, but also in kidney and vascular cells [193]. While all Noxes are present in mice, rats and man, the rodent genome does not contain the *nox5* gene [194]. Unlike other vascular Noxes, Nox5 possesses an amino-terminal calmodulin-like domain with four binding sites for Ca^{2+} (EF hands) and unique to Nox5 is its lack of requirement for p22phox or other subunits for its activation. Nox5 is directly regulated by intracellular Ca^{2+} ($[Ca^{2+}]_i$), the binding of which induces a conformational change leading to enhanced ROS formation [193]. The biological significance of vascular Nox5 is unknown, although it has been implicated in cell proliferation, angiogenesis and migration and in oxidative damage in atherosclerosis [194]. Vascular Nox5 is activated by thrombin, PDGF, Ang II and ET-1 [195] and its expression is regulated by the Ca^{2+}-sensitive transcription factor CREB. Increased Nox5 expression has been demonstrated in coronary arteries from patients with coronary artery disease [196].

·   ·   ·   ·   ·

CHAPTER 4

Interactions between Mitochondria and Noxs

Mitochondrial enzymes and NADPH oxidase are activated in pathological processes at the same time, and it has been suggested that cross-talk between mitochondria and Noxs, particularly Nox4, may be important in dysregulated ROS formation [197]. Such cross talk may be important in oxidative stress associated with nitroglycerin-triggered vascular dysfunction, myocardial infarction, cardiac failure and vascular remodeling [197, 198]. Mitochondrial:Nox interaction may also regulate vascular oxygen-sensing mechanisms [199]. Mechanisms linking these ROS-generating systems remain involve in ERK1/2 [200], the mitochondrial permeability transition pore and ATP-sensitive potassium channels [201].

. . . .

CHAPTER 5

Distribution of Noxes in the Vascular Wall

The three major cell types of the vascular wall including endothelial cells, smooth muscle cells and adventitial fibroblasts, all possess functionally active Nox isoforms [201–203] (Figure 5). In pathological conditions associated with vascular injury, such as atherosclerosis, diabetes and hypertension, macrophages and leukocytes invade the vessel and become resident cells in the vascular media [204]. These cells are rich in NAD(P)H oxidase and may also contribute to vascular ROS generation. Endothelial cells express mRNA and protein for Nox2, Nox4 and associated regulatory proteins p22phox, p47phox and p67phox and play a role in endothelial cell biology [205]. Nox2 is the major source of ROS in endothelial cells under basal conditions, and in pathological conditions, Nox1 and Nox4 may be upregulated [206, 207]. Nox2, Nox4 and Nox5 appear to localize primarily in the perinuclear area associated with membranes on the endoplasmic reticulum and nucleus although Nox2 is also found in the plasma membrane within cholesterol-enriched domains, which may serve as signaling platforms for ROS generation in vascular disease [208–211]. Vascular smooth muscle cells possess Nox2 (in human resistance arteries) and Nox4, which are major sources of ROS. Nox1, present in low concentrations in basal states, is upregulated in disease. Adventitial fibroblasts also possess Noxes (Nox2, Nox4), important in adventitial ROS formation.

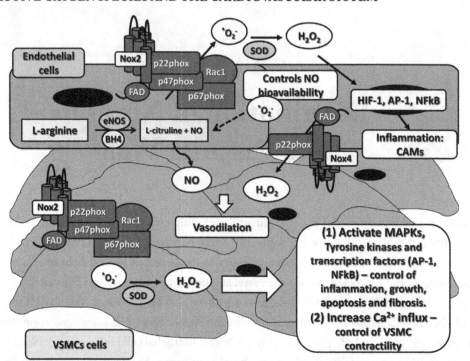

FIGURE 5: Physiological effects of ROS in the vasculature. Superoxide and other ROS are also important for the regulation of a series of physiological responses in the vessels. ROS regulate vascular tone through vasodilation, mediated by hydrogen peroxide, and control of NO bioavailability, mediated through superoxide. ROS is also important as a sensor of the oxygen levels in the vasculature, through regulation of the expression of HIF-1. ROS regulates the activity of transcription factors and production of pro-inflammatory proteins, such as adhesion molecules and cytokines in the endothelium. In VSMCs, ROS plays an important role in the regulation of calcium influx and cell contractility and migration, as well as, in the regulation of signaling cascades involved in growth, fibrotic and inflammatory responses.

· · · ·

CHAPTER 6

Regulation of Noxes

How the NAD(P)H oxidase subunits interact in cardiovascular cells and how they generate $\bullet O_2^-$ is still unclear. All Noxes, except Nox5, appear to have an obligatory need for p22phox [212–214]. Whereas Nox2 requires p47phox and p67phox for its activity, Nox1 may interact with homologues of p47phox (NAD(P)H oxidase organizer 1 (NOXO1)) and p67phox (NAD(P)H oxidase activator 1 (NOXA1)) [215, 216]. Oxidase activation involves Rac translocation, phosphorylation of p47phox and its translocation, possibly with p67phox and p47phox association with cytochrome b558. Nox2 and Nox 4 are constitutively active. However, induction of Nox mRNA expression is observed in response to physical stimuli, (shear stress, pressure), growth factors (platelet-derived growth factor, epidermal growth factor and transforming growth factor β), cytokines (tumor necrosis factor-α, interleukin-1 and platelet aggregation factor), mechanical forces (cyclic stretch, laminar and oscillatory shear stress), metabolic factors (hyperglycemia, hyperinsulinemia, free fatty acids, advanced glycation end products (AGE) and G protein-coupled receptor agonists (serotonin, thrombin, bradykinin, endothelin and Ang II) [217–222]. Ang II is an important and potent regulator of cardiovascular NAD(P)H oxidase, which activates NAD(P)H oxidase via AT_1 receptors through stimulation of signaling pathways involving c-Src p21^{Ras}, PKC, PLD and PLA$_2$ [223–226]. Ang II also influences NAD(P)H oxidase activation through transcriptional regulation of oxidase subunits.

· · · ·

CHAPTER 7

Protecting Against Oxidative Stress—Antioxidant Defenses

Antioxidants are substances that at low concentrations prevent or inhibit oxidation of oxidizable biomolecules, such as DNA, lipids and proteins. Enzymatic and nonenzymatic systems have evolved to protect against injurious oxidative stress. Major enzymatic antioxidants are SOD, catalase, glutathione peroxidases, thioredoxin and peroxiredoxin [227–230] (Figure 6). Non-enzymatic

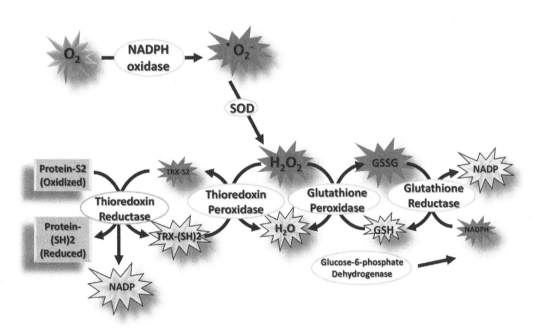

FIGURE 6: Antioxidant systems in the vasculature. SOD catalyzes the dismutation of superoxide into hydrogen peroxide, which is converted to water by catalase. Glutathione peroxidase plays a role in hydrogen peroxide reduction to water. During this process, oxidized glutathione is formed and is rapidly recycled by glutathione reductase, a process that consumes NADPH. Thioredoxin reductase participates in thiol-dependent cellular reductive processes and activation of proteins that were oxidized and inactivated by ROS. Thioredoxin peroxidase is responsible for the degradation of hydrogen peroxidase into water.

antioxidants include ascorbate, tocopherols, glutathione, billirubin and uric acid and scavenge OH• and other free radicals [231].

7.1   ENZYMATIC ANTIOXIDANTS

SOD catalyzes the dismutation of $\cdot O_2^-$ into H_2O_2 and O_2. Extracellular SOD, the major vascular SOD, is produced and secreted by vascular smooth muscle cells and binds to glycosaminoglycans in the vascular extracellular matrix and regulates oxidant status in the vascular interstitium [227, 230]. Reduced glutathione plays a major role in the regulation of the intracellular redox state of vascular cells by providing reducing equivalents for many biochemical pathways [231–233]. Glutathione peroxidase (GPX) reduces H_2O_2 and lipid peroxides to water and lipid alcohols, respectively, and in turn oxidizes glutathione to glutathione disulfide [231]. Oxidized glutathione (GSSG) can be recycled by glutathione reductase to reduced GSH utilizing NADPH as a substrate or it can be exported from the cell via active transport by the multidrug resistance protein 1 (MRP1) [234, 235]. Hypertension induced by DOCA-salt or Ang II was attenuated in MRP–/– mice, and vascular glutathione flux was blunted in MRP1–/– mice allowing recycling of GSSG to reduced glutathione and promoting increased intracellular antioxidant capacity [234, 235]. These findings suggest that MRP1 inhibition may protect against oxidant stress by preventing loss of glutathione from vascular cells, thereby improving endothelial function and attenuating development of hypertension. Catalase is an intracellular antioxidant enzyme that is mainly located in cellular peroxisomes and catalyzes the reaction of H_2O_2 to water and O_2 [236]. Catalase is very effective in high-level oxidative stress and protects cells from H_2O_2 produced within the cell. The enzyme is especially important in the case of limited glutathione content or reduced GPX activity. Thioredoxin reductase participates in thiol-dependent cellular reductive processes [237–239]. Low antioxidant bioavailability promotes cellular oxidative stress and has been implicated in cardiovascular and renal oxidative damage associated with hypertension [227]. Activity of SOD, catalase and GSH peroxidase is lower and the GSSG/GSH is higher in plasma and circulating cells from hypertensive patients than normotensive subjects [240]. In mice deficient in EC-SOD and in rats in which GSH synthesis is inhibited, blood pressure is significantly elevated, demonstrating that reduced antioxidant capacity is associated with elevated blood pressure [241, 242]. Failure to upregulate antioxidant genes and reduced antioxidant capacity are also associated with age-accelerated atherosclerosis [243].

7.2   NON-ENZYMATIC ANTIOXIDANTS

The major biological non-enzymatic systems include vitamin E, vitamin C and glutathione [244, 245]. Vitamin E (tocopherols and tocotrienols) is the primary lipid-soluble chain-breaking antioxidant in the body. Vitamin E traps peroxy radicals in cell membranes and other lipid microenvironments. Vitamin C (ascorbic acid) is a water-soluble antioxidant involved in the reduction of radicals by recycling radicals produced by oxidation of vitamin E. Vitamins E and C, at high concentrations, can function as pro-oxidants causing cell damage [246]. Some of the negative clinical outcomes examining cardiovascular effects of antioxidant vitamins have been attributed to the pro-oxidant capacity of vitamins E and C. Glutathione is an important intracellular antioxidant. The cysteine residue on glutathione provides an exposed free sulphydryl group that is highly reactive, providing a rich radical target. Reaction with free radicals oxidizes glutathione but the reduced form is regenerated in a redox cycle involving glutathione reductase and NADPH as an electron acceptor [247].

Several other micromolecules such as bilirubin, uric acid, carotenoids among others can function as antioxidants [248]. Carotenoids are naturally occurring lipophilic compounds with β-carotene being the most abundant. They have many conjugated double bonds which are responsible for their antioxidant properties. Bilirubin inhibits lipid oxidation and oxygen radical formation and may act as a physiological antioxidant providing protection against cardiovascular disease. Many studies have demonstrated an inverse association between plasma bilirubin levels and cardiovascular-related risk factors such as diabetes, metabolic syndrome, hypertension and body mass index and as such bilirubin has been suggested as a potential clinical biomarker of cardiovascular disease [249, 250]. Individuals with Gilberts syndrome, who have decreased hepatic bilirubin UDP-glucuronosyltransferase activity, decreased bilirubin clearance, and increased serum unconjugated bilirubin concentrations have a lower prevalence of cardiovascular disease, possibly related to the antioxidant capacity of increased bilirubin [251]. Because of the protective effects of antioxidants, there has been much interest in developing synthetic and natural antioxidants as therapeutic agents to prevent and/or treat patients with cardiovascular disease. However, to date, there is still little definitive proof that such strategies are effective.

·   ·   ·   ·

CHAPTER 8

How Does Oxidative Stress Cause Disease?

Oxidative stress, as proposed by H. Sies, is a change in the prooxidant/antioxidant balance of a biologic system in favour of pro-oxidation [252, 253]. This occurs when the generation of ROS and RNS is greater than the antioxidant capacity, resulting in an increase in the bioavailability of reactive molecules. Consequences of an oxidative burden include oxidative damage of biomolecules, including DNA, proteins and lipids and altered signal transduction. Many chronic diseases, including neurodegenerative, cardiovascular and chronic kidney disease, diabetes, hypertension and cancer have been associated, directly or indirectly, with oxidative injury, induced in large part by redox-sensitive inflammation, hypertrophy, proliferation, apoptosis, migration and fibrosis. In the vascular system, these responses underlie structural remodeling, endothelial dysfunction, inflammation, thrombosis and vasoconstriction, which are associated with hypertension, atherosclerosis, obesity, diabetes, stroke, kidney disease and cardiac failure.

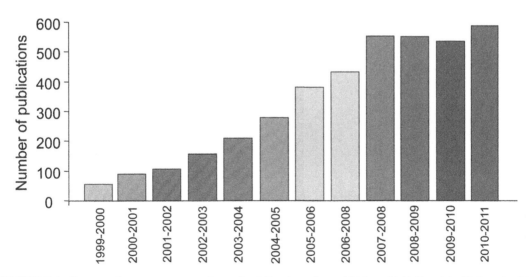

FIGURE 7: Bar graph represents number of publications from 1999 to 2011 listed in Pubmed with "oxidative stress" and "hypertension" as key words.

The interest in ROS and disease is evidenced by the increasing number of publications related to oxidative stress, with over 1% of all papers searched in Pubmed since 2005 as having the term "oxidative stress" as a key word [254]. Similar trends have been observed for hypertension, where the number of publications with "hypertension" and "oxidative stress" as keywords has increased more than 10-fold since the 1990s (Figure 7). Although there is abundant data supporting oxidative injury in human disease, only few pathologies have been directly linked to oxidative stress. These include chronic deprivation of selenium and reduced intake of vitamin E due to abnormal intestinal fat absorption, pathologies associated with altered metal homeostasis, such as Wilsons disease, chronic granulomatous disease and amyotrophic lateral sclerosis [254–257]. In other diseases, the direct association between oxidative stress and the pathogenic process still awaits confirmation.

. . . .

CHAPTER 9

Oxidative Stress and Hypertension

9.1 PRODUCTION OF ROS IN THE CARDIOVASCULAR AND RENAL SYSTEMS IN HYPERTENSION

ROS have been implicated in the regulation of vascular tone by modulating vasodilation directly (H_2O_2 may have vasodilator actions) or indirectly by decreasing NO bioavailability through quenching by $\bullet O_2^-$ to form $ONOO^-$ [258, 259]. ROS, through the regulation of hypoxia-inducible factor-1 (HIF-1), are also important in O_2 sensing [260], which is essential for maintaining normal O_2 homeostasis. In pathological conditions, ROS are involved in inflammation, endothelial dysfunction, cell proliferation, migration and activation, extracellular matrix deposition, fibrosis, angiogenesis and vascular remodeling. These effects are mediated through redox-sensitive regulation of multiple signaling molecules and second messengers including mitogen-activated protein (MAP) kinases, protein tyrosine phosphatases, tyrosine kinases, pro-inflammatory genes, ion channels and Ca^{2+} (Figure 8) [261–265].

Mechanisms whereby ROS influence the development of hypertension involve oxidative damage of multiple systems including the heart, kidneys, central and peripheral nervous system and vasculature. In pathological conditions, ROS are involved in inflammation, endothelial dysfunction, cell proliferation, migration and activation, extracellular matrix deposition and fibrosis. These effects are mediated through redox-sensitive signaling pathways including MAPK, PTP, tyrosine kinases, pro-inflammatory genes, ion channels and Ca^{2+} [266, 267].

Changes in cardiovascular function and structure probably relate to oxidative stress-induced endothelial dysfunction, reduced vasodilation, increased contraction, vascular inflammation and structural remodeling causing increased peripheral resistance and elevated blood pressure (Figures 9, 10) [268]. Centrally produced ROS by NADPH oxidase in the hypothalamic and circumventricular organs are implicated in central control of hypertension, in part through sympathetic outflow [269]. Oxidative stress in the central nervous system has been implicated in the pathogenesis of stroke, an important complication of hypertension (Figure 11).

The kidney, and particularly the renal medullary circulation, plays a fundamental role in modulating long-term blood pressure control and fluid balance (Figure 12). ROS are important regulators of medullary blood flow. Elevation of $\bullet O_2^-$ or reduction of NO in the renal medulla

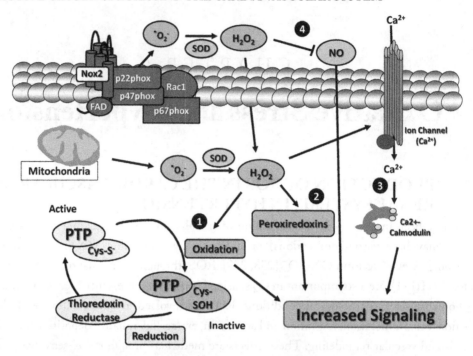

FIGURE 8: Mechanisms whereby ROS induces activation of signaling proteins. It is well known that ROS can increase activation of redox-sensitive proteins important for signaling and regulation of cell function. The mechanisms involved in this process are still not fully characterized. However, ROS can increase cell signaling through oxidation of protein tyrosine phosphatases (PTPs), leading to inactivation of PTPS. Peroxiredoxins may also play a role in ROS-induced signaling. ROS also controls the activity of ion channels (i.e. calcium channels), leading to an increase in ion influx/efflux and activation of the respective signaling. Another mechanism related to ROS is the interactions with other molecules. ROS can increase and/or decrease the bioavailability of certain molecules, i.e., NO, preventing their effects.

decreases medullary blood flow and Na^+ excretion, resulting in sustained hypertension [270]. Oxidative stress within the renal medulla makes the kidney functionally more vulnerable to effects of Ang II and salt and promotes renal dysfunction [271]. Superoxide and H_2O_2 augment afferent arterial tone and reactivity and enhance renal vascular resistance. Renal ROS influence glomerular filtration rate, tubuloglomerular feedback response and Na^+ transport. NO inhibits absorption of NaCl in the thick ascending limb, whereas $\bullet O_2^-$ enhances NaCl reabsorption. Moreover, Ang II and oxidative stress promote mesangial cell proliferation, mesangial matrix accumulation and podocyte injury, which are hallmarks of glomerulonephritis and diabetic nephropathy [272].

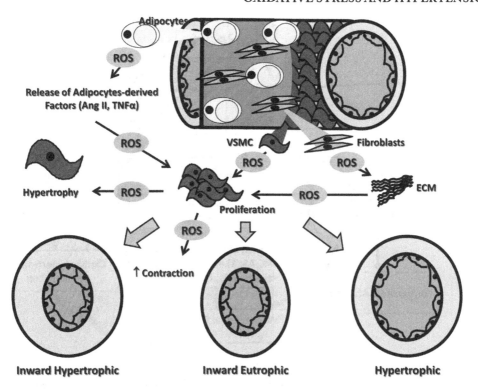

FIGURE 9: Role of ROS in vascular remodeling. Cells from different layers of the vasculature are capable of generating ROS. Adipocytes in the adventitia produce adipocyte-derived factors, which induce proliferation and hypertrophy in VSMCs and increase extracellular matrix deposition by fibroblasts. Vasoactive peptides, from endothelial cells or the circulation, also control pro-proliferative, pro-inflammatory, pro-fibrotic and contractile responses in VSMCs. All these processes are redox-sensitive and lead to remodeling. ROS are involved in the many different types of vascular remodeling, observed in hypertension, diabetes and stroke.

Increased renal ROS production induces upregulation of pro-inflammatory genes, such as hypoxia-inducible factor and AP-1, implicated in inflammation, fibrosis and sclerosis in hypertension-associated kidney damage. Renal ROS generation is especially important in severe and salt-dependent forms of hypertension such as Dahl salt-sensitive rats, DOCA-salt rats and stroke-prone SHR (SHR-SP) [273]. In these models, blood pressure lowering effects of the SOD mimetic tempol were associated with reduced renal excretion of 8-isoprostane PGF2α, decreased vascular resistance, increased GFR and enhanced diuresis and natriuresis [274].

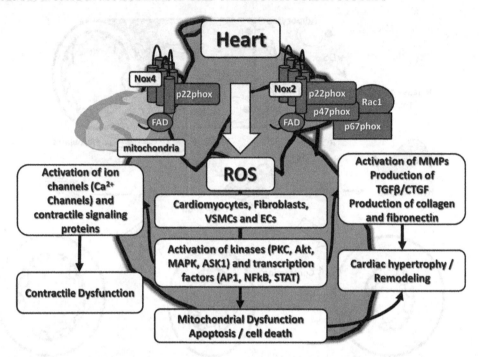

FIGURE 10: ROS effects in the heart. ROS are produced in the heart mainly by Nox2, Nox4 and the mitochondria. The role of Nox4 in heart tissue is still unclear and controversial. It seems that Nox4 can co-localize with the mitochondria in cardiomyocytes, where ROS generated from this complex is deleterious. However, other studies have suggested a protective role of Nox4 in cardiomyocytes and coronary endothelial cells. ROS generated in cardiac tissue may lead to increase in activation of signaling proteins, and consequently, to cell death, production of growth factors and/or inflammatory cytokines, extracellular matrix remodeling and activation of contractile proteins. All these cellular responses will then play a role in cardiac hypertrophy/remodeling and dysfunction.

Renal oxidative stress is also important in kidney disease associated with Ang II-dependent forms of hypertension. In TG(mRen2)27 (Ren2) transgenic rats, which overexpress the mouse renin gene, AT_1R blockade reduced blood pressure and tissue oxidative stress, improved glomerular filtration barrier integrity and prevented albuminuria, thereby slowing progression of kidney disease [275].

A major source of renal $\bullet O_2^-$ is NAD(P)H oxidase. All of the major Noxs and NAD(P)H oxidase subunits are present in the rat renal cortex and outer medulla [276, 277]. These proteins are expressed primarily in renal arterioles, glomeruli and the distal nephron and are abundantly expressed at the luminal border of macula densa cells and in podocytes. In SHR, Ang II-induced

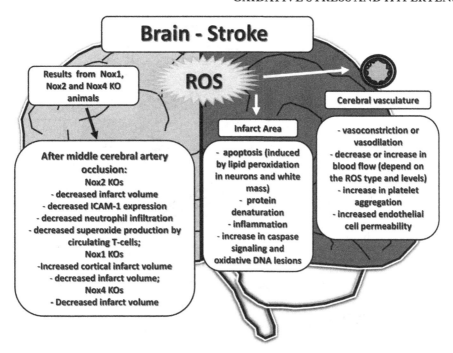

FIGURE 11: Role of ROS in stroke. Oxidative stress plays an important role in stroke. ROS is able to induce vasoconstriction or vasodilation in arteries from the brain, depending on the free radical that is formed. These changes in ROS levels lead to: insufficient oxygenation of the cerebral tissue, increase in platelet aggregation and endothelial cell permeability. Increased apoptosis of neurons, protein denaturation, DNA damage and inflammation, induced by free radicals, are observed in infarct areas. In animal models of cerebral ischemia, depletion of Nox2, Nox1 and Nox4, allowed researchers to link Nox-derived ROS to infract volume, pro-inflammatory molecules expression and inflammatory responses.

hypertension and in salt-sensitive hypertension, renal expression of NAD(P)H oxidase subunits is increased and activity of Nox is enhanced.

9.2   ROS AND AUTONOMIC OUTFLOW AND HYPERTENSION

In physiological conditions, autonomic outflow is regulated by the balance between sympathoinhibitory effects of NO and the sympathostimulatory effects of O_2^-. In pathological conditions, increased ROS generation, due in part to hyperactivation of NADPH oxidase and mitochondrial oxidases, stimulates central sympathetic outflow promoting sympathetic hyperactivity, an effect that is normalized by antioxidant therapy in experimental hypertension and cardiac failure [278, 279]. Centrally produced ROS by NAD(P)H oxidase in the hypothalamic and circumventricular organs

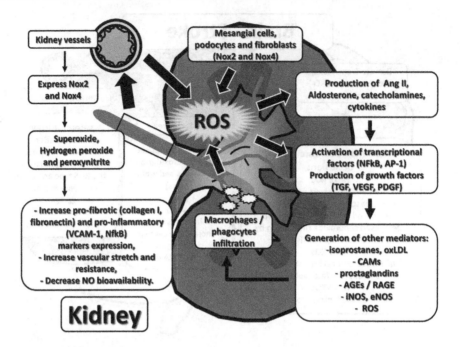

FIGURE 12: ROS effects in the kidney. ROS can be formed by infiltrating phagocytes, vascular cells, podocytes, fibroblasts, epithelial cells and mesangial cells in the kidney. Once generated, ROS regulates: production and activity of vasoactive peptides, hormones and cytokines; activation of transcriptional factors; production of growth factors; expression of adhesion molecules, prostaglandins, iNOS and eNOS; production of extracellular matrix, vascular tone and NO bioavailability.

are implicated in central control of hypertension, in part through sympathetic outflow [280]. Injection of SOD into the rostral ventrolateral medulla (RVLM) decreased sympathetic nerve activity in swine and in rat experimental models of hypertension, RVLM ROS levels are increased [278–280]. Regulation of sympathetic nerve activity by central ROS involves Ang II/AT1R [280–282]. Rabbits with cardiac failure exhibit increased renal sympathetic nerve activity and arterial baroreflex function, effects associated with enhanced NADPH oxidase-derived O_2^- production in the RVLM and upregulation of Ang II/AT1R. These effects were restored to near normal by tempol, the SOD mimetic or by apocynin, NADPH oxidase inhibitor [280]. Mechanisms whereby O_2^- mediates an increase in sympathetic outflow by central neurons are unclear but activation of specific populations of neurons through alterations in potassium and calcium channels may be important. In addition, reductions in the sympathoinhibitory influence of NO because of increased scavenging of NO to form ONOO- and a reduction in nNOS activity may contribute to sympathoexcitation [283].

Reactive oxygen species play an important role in regulating autonomic balance. Oxidative stress is implicated in sympathetic hyperactivity, neuronal apoptosis and death. These findings have evoked considerable interest because of the possibilities that oxidative stress in the nervous system may contribute to disorders associated with autonomic dysfunction and to neurological and psychiatric diseases. Hence, therapies targeted to decrease ROS generation and/or strategies to increase NO availability in the central and peripheral nervous system may be useful in minimizing neural injury and thereby prevent oxidative stress-related neurological diseases.

9.3   OXIDATIVE STRESS IN EXPERIMENTAL HYPERTENSION

The relationship between oxidative stress and increased blood pressure has been demonstrated in many models of hypertension. Increased ROS formation precedes development of hypertension in SHR and is implicated in fetal programming and development of hypertension later in life, supporting the important role of ROS in the genesis and maintenance of hypertension [284, 285]. Markers of oxidative stress, such as TBARS, and $F_{2\alpha}$-isoprostanes, tissue concentrations of $\bullet O_2^-$ and H_2O_2 and activation of NAD(P)H oxidase and xanthine oxidase are increased, whereas levels of NO and antioxidant enzymes are reduced in experimental hypertension [286–289].

Ang II-dependent hypertension is particularly sensitive to NAD(P)H oxidase-derived ROS. In rats and mice made hypertensive by Ang II infusion, expression of NAD(P)H oxidase subunits (Nox1, Nox2, Nox4, p22phox), oxidase activity and generation of ROS are increased [290–294]. To support a role for NAD(P)H oxidase-derived ROS generation in the pathogenesis of Ang II-induced hypertension, various mouse models with altered NAD(P)H oxidase subunit expression have been studied [295–298]. In p47phox knockout mice and in gp91phox (Nox2) knockout mice, Ang II infusion fails to induce hypertension and these animals do not show the same increases in $\bullet O_2^-$ production, vascular hypertrophy and endothelial dysfunction observed in Ang II-infused wild-type mice [298, 299]. In Ang II-infused mice treated with siRNA targeted to renal p22phox, renal NAD(P)H oxidase activity was blunted, ROS formation was reduced and blood pressure elevation was attenuated, suggesting that p22phox is required for Ang II-induced oxidative stress and hypertension [300]. On the other hand, overexpression of vascular p22phox was associated with increased oxidative stress and vascular dysfunction, but no significant increase in blood pressure [301]. Treatment with apocynin or diphenylene iodinium, non-specific pharmacological inhibitors of NAD(P)H oxidase, or gp91dstat, a novel specific inhibitor of NAD(P)H oxidase, reduced vascular $\bullet O_2^-$ production, prevented cardiovascular remodeling and attenuated development of hypertension in Ang II-treated mice [302–304]. Nox1-deficient mice have reduced vascular $\bullet O_2^-$ production and blood pressure elevation in response to Ang II is blunted [305, 306], whereas in transgenic mice in which Nox1 is overxpressed in the vascular wall, Ang II-mediated vascular hypertrophy and blood pressure elevation are enhanced [307]. In most of these models, Ang II was infused for a short time period (1–3 weeks), inducing an acute hypertensive response. In a model of chronic Ang II-dependent hypertension, where we crossed transgenic mice expressing human renin (which exhibit an Ang II-sensitive hypertensive phenotype), with Nox2–/– or Nox1–/– mice, development of hypertension was not prevented even though oxidative stress was reduced, suggesting that Noxes may be more important in acute than in chronic hypertension [173, 179].

There is also evidence for ROS involvement in the pathogenesis of hypertension independent of direct Ang II actions [308–318]. In SHR, vascular, renal and cardiac $\bullet O_2^-$ production is enhanced compared with normotensive controls [309–312]. In stroke-prone SHR, aortic expression of Nox1 and Nox4 is significantly increased compared with WKY [313]. In DOCA salt-induced mineralocorticoid hypertension vascular $\bullet O_2^-$ production involving elevated NAD(P)H oxidase activity, uncoupling of endothelial NOS and mitochondrial sources, in part through the endothelin-1 (ET-1)/ETA receptor pathway, is increased [314–317]. Infusion of ET-1 increases NAD(P)H oxidase-dependent $\bullet O_2^-$ production; however, preventing this increase in ROS generation does not inhibit development of hypertension in these animals [318]. Overexpression of human ET-1 in mice also induces vascular remodeling and impairs endothelial function, via activation of NAD(P)H oxidase [319]. To further support a role for oxidative stress in hypertension, many studies have shown that treatment with antioxidant vitamins, SOD, such as tempol (4-hydroxy-2,2,6,6-tetramethyl piperidinoxyl), free radical scavengers or tetrahydrobiopterin (BH4) attenuate or prevent development of hypertension and associated target organ damage [320–322].

9.4 OXIDATIVE STRESS AND CLINICAL HYPERTENSION

Although studies in humans have not been as convincing as those in experimental models, there is evidence that oxidative stress is increased in patients with essential hypertension, renovascular hypertension, malignant hypertension, salt-sensitive hypertension, cyclosporine-induced hypertension and preeclampsia [323–325]. These findings are based, in general, on increased levels of plasma thiobarbituric acid-reactive substances and 8-epi-isoprostanes, biomarkers of lipid peroxidation and oxidative stress [326–328]. Polymorphonuclear leukocyte- and platelet-derived $\bullet O_2^-$, which also participate in vascular oxidative stress and inflammation, are increased in hypertensive patients [329, 330].

Hypertensive patients exhibit a significantly higher production of plasma H_2O_2 than normotensive subjects [331]. Additionally, normotensive subjects with a family history of hypertension have greater H_2O_2 production than blood pressure-matched normotensives without a family history of hypertension, suggesting that there may be a genetic component that leads to elevated production of hydrogen peroxide [331, 332]. Plasma levels of asymmetric dimethylarginine (ADMA) (eNOS inhibitor) and the lipid peroxidation product of linoleic acid, 13-hydroxyoctadecadienoic acid (HODE), a marker of ROS production, were inversely correlated with microvascular endothelial dysfunction and elevated blood pressure in hypertensive patients [333].

We showed that ROS production is increased in vascular smooth muscle cells from resistance arteries of hypertensive patients and that this is associated with upregulation of vascular NAD(P)H oxidase [334, 335]. The importance of this oxidase in oxidative stress in human cardiovascular disease is supported by studies from Zalba and colleagues who demonstrated that polymorphisms in NAD(P)H oxidase subunits are associated with increased atherosclerosis and hypertension [336]. In particular, the −930(A/G) polymorphism in the p22(phox) promoter may be a novel genetic marker associated with hypertension [337]. The C242T CYBA polymorphism is associated with essential hypertension and hypertensive patients carrying the CC genotype of this polymorphism exhibit features of NAD(P)H oxidase-mediated oxidative stress and endothelial damage and are prone to cerebrovascular disease [337, 338]. In a Japanese population, the G(−930)A polymorphism of CYBA was confirmed to be important in the pathogenesis of hypertension [339]. Polymorphisms −337GA and 565+64CT of xanthine oxidase gene have been shown to be related to blood pressure and oxidative stress in hypertension, further supporting a role for xanthine oxidase in hypertension.

In addition to excess ROS generation, decreased antioxidant defense mechanisms contribute to oxidative stress in patients with hypertension. Hypertensive patients have reduced activity and decreased content of antioxidant enzymes, including SOD, glutathione peroxidase and catalase [340–342]. Decreased levels of antioxidant vitamins A, C and E have been demonstrated in newly diagnosed, untreated hypertensive patients compared with normotensive controls [342]. Moreover, SOD activity has been demonstrated to correlate inversely with blood pressure in patients with hypertension [342]. Antioxidant vitamins reduced blood pressure and arterial stiffness in patients with diabetes [343], but had no effect in postmenopausal women or in healthy subjects [344]. In patients with white coat hypertension, serum protein carbonyl (PCO, indicating protein oxidation) was increased and endogenous antioxidant proteins (protein thiol, SOD, glutathione) were decreased compared with normotensive individuals, further supporting a relationship between oxidative stress and hypertension [345].

·   ·   ·   ·

CHAPTER 10

Antioxidant Therapy and Human Hypertension

The potential of antioxidants in treating conditions associated with oxidative stress is supported by experimental investigations, observational findings, small clinical studies and epidemiological data [346, 347]. However, findings are inconsistent and clinical trial data are inconclusive [348, 349]. Many large trials have been published regarding antioxidant vitamin effects on risks of cardiovascular disease, including the Cambridge Heart Antioxidant Study (CHAOS; 2002 patients); Alpha Tocopherol, Beta-Carotene cancer prevention study (ATBC; 27 271 males); Gruppo Italiano per lo Studio della Sopravvivenza nell'Infarto Miocardico (GISSI)-Prevenzione trial (3658 patients); Heart Outcomes Prevention Evaluation (HOPE) study (2545 subjects); Medical Research Council/ British Heart Foundation (MRC/BHF) heart protection study (20 536 adults); Primary Prevention Project (PPP; 4495 patients); and the Antioxidant Supplementation in Atherosclerosis Prevention (ASAP) study (520 subjects) [348, 349]. In the HOPE-TOO study, which was a follow-up of a subset of the original HOPE trial (Heart Outcomes Prevention Evaluation), patients taking 400 IU vitamin E showed increased incidence of heart failure [348–350]. Except for the ASAP study, which demonstrated that 6-year supplementation of daily vitamin E and slow-release vitamin C reduced progression of carotid atherosclerosis, the other studies failed to demonstrate significant beneficial effects of antioxidants on BP or on cardiovascular end points. Thus, overall results of clinical trials have been negative [351].

Unlike the large multicenter trials, smaller clinical studies have shown positive responses in hypertensive patients treated with antioxidants, either in combination (zinc, ascorbic acid, α-tocopherol, β-carotene) or as mono-therapy (vitamin C or vitamin E). This has been particularly true for vitamin C. Most studies demonstrated an inverse relationship between plasma ascorbate levels and blood pressure in both normotensive and hypertensive populations [352]. In the SU.VI.MAX study, a decreasing trend was observed with vitamin C levels and risk of hypertension in women but not in men [353]. Vitamin C supplementation is associated with reduced blood pressure in hypertensive patients with systolic blood pressure falling by 3.6–17.8 mm Hg for each 50 µmol/L increase in plasma ascorbate [354, 355]. However, Ward et al. found that 6-week treatment with vitamin C and grape seed polyphenols was associated with a paradoxical increase in ambulatory blood pressure in hypertensive patients [356]. This was not attributed to increased oxidative stress.

Human studies of vitamin E (400–1000 IU/day) have demonstrated beneficial effects in improving insulin sensitivity, lowering serum glucose levels, increasing intracellular Mg^{2+}, inhibiting thromboxane effects and reducing vascular resistance [352, 357, 358]. Data from the 1946 British Birth Cohort reported that low vitamin E intake during childhood and adulthood was a good predictor of hypertension at age 43 years [358]. However, reductions in blood pressure in hypertensive subjects treated with vitamin E have been inconsistent [240, 352]. Similar trends have been observed in preeclampsia, where early studies suggested that vitamins C and E may prevent against preeclampsia in high risk patients [359, 360], whereas more recent evidence indicates that supplementation with vitamins C and E during pregnancy does not reduce the risk of preeclampsia in nulliparous women [361–363]. If vitamin E does in fact have an antihypertensive effect, it is probably small and may be limited to untreated patients or those with vascular disease or other concomitant diseases, such as diabetes [361–364].

In general, results of clinical studies investigating antioxidant effects have been disappointing given the consistent and promising findings from experimental investigations, clinical observations and epidemiological data [351]. Possible reasons relate to (1) type of antioxidants used, (2) patient cohorts included in trials and (3) the trial design itself. With respect to antioxidants, it is possible that agents examined were ineffective and non-specific and that dosing regimens and duration of therapy were insufficient. For example, vitamins C and E may have pro-oxidant properties with harmful and deleterious interactions. It is also possible that orally administered antioxidants may be inaccessible to the source of free radicals, particularly if ROS are generated in intracellular compartments and organelles [351, 365]. Furthermore, antioxidant vitamins do not scavenge H_2O_2, which may be more important than $\bullet O_2^-$ in cardiovascular disease. Another factor of importance is that antioxidants do not inhibit ROS production. Regarding cohorts included in large trials, most subjects had significant cardiovascular disease, in which case damaging effects of oxidative stress may be irreversible. Another confounding factor is that most of the enrolled subjects were taking aspirin prophylactically. Since aspirin has intrinsic antioxidant properties [366] additional antioxidant therapy may be ineffective. Moreover, in patients studied in whom negative results were obtained, it was never proven that these individuals did in fact have increased oxidative stress. To date, there are no large clinical trials in which patients were recruited based on evidence of elevated ROS formation. Also, none of the large clinical trials were designed to examine effects of antioxidants specifically on blood pressure.

. . . .

CHAPTER 11

NADPH Oxidase and Nox Isoforms as Therapeutic Targets—Clinical Potential

Based on experimental evidence, NADPH oxidase subunits and Nox isoforms are potential thera-peutic targets for cardiovascular disease and hypertension. Because of this, there has been enormous interest in the development of agents that inhibit NADPH oxidases in an isoform-specific manner [367, 368]. Different strategies have been employed, including small molecule inhibitors, peptide NADPH oxidase inhibitors and siRNAS. Several compounds have been registered as NADPH oxidase inhibitors in the patent literature [367, 368]. However, none have gone through clinical trials and some have not yet completed preclinical studies. To date, two different classes of com-pounds have been claimed as potent and orally active bioavailable NADPH oxidase inhibitors: pyr-azolopyridines (GKT136901 and GKT137831) and triazolopyrimidine derivatives (VAS2870 and VAS3947). Although the mechanisms of inhibition have not yet been clarified, GKT compounds may act as competitive substrate inhibitors, since structurally, they resemble NADPH. Although much research is still needed to confirm the clinical use of NADPH oxidase inhibitors in humans, these drugs hold promise in the management of patients with Nox-associated pathologies.

CHAPTER

NADPH Oxidases and Nox Isoforms as Therapeutic Targets—Clinical Potential

CHAPTER 12

Other Strategies to Reduce Oxidative Stress

Theoretically, agents that reduce oxidant formation should be more efficacious than non-specific, inefficient antioxidant vitamin scavengers. This is based on experimental evidence where it has been demonstrated that inhibition of NAD(P)H oxidase-mediated $\bullet O_2^-$ generation, using pharmacological and gene-targeted strategies, leads to regression of vascular remodeling, improved endothelial function and lowering of blood pressure [369–373]. In fact, vascular NAD(P)H oxidase, specifically gp91phox (nox2) homologues may be novel therapeutic targets for vascular disease [369–372]. Harrison and colleagues [234] proposed a new strategy to increase antioxidant capacity without the use of exogenous antioxidants. They suggest that drugs that selectively inhibit MRP1 would prevent cellular glutathione loss and thereby protect against oxidative damage, endothelial dysfunction and hypertension [234]. Another interesting approach is targeting glucose-6-phosphate dehydrogenase (G6PD), which is a source of NADPH, the substrate for NAD(P)H oxidase [373]. Inhibition of G6PD has been shown to ameliorate development of pulmonary hypertension, possibly through decreased oxidative stress. To date, only investigational G6PD inhibitors are available.

In view of current data and the lack of evidence to prove the benefits from use of antioxidants to prevent cardiovascular disease [71, 373], antioxidant supplementation is not recommended for the prevention or treatment of hypertension. However, most therapeutic guidelines suggest that the general population consumes a diet emphasizing antioxidant-rich fruits and vegetables and whole grains [374–377]. Another important lifestyle modification that may have cardiovascular protective and blood pressure lowering effects by reducing oxidative stress is exercise. In experimental models of hypertension and in human patients with coronary artery disease, exercise reduced vascular NAD(P)H oxidase activity and ROS production, ameliorated vascular injury and reduced blood pressure [378–385].

Some of the beneficial effects of classical antihypertensive agents such as ß-adrenergic blockers, ACE inhibitors, AT_1 receptor antagonists, and Ca^{2+} channel blockers may be mediated, in part, by decreasing vascular oxidative stress [381, 382, 385]. These effects have been attributed to direct inhibition of NAD(P)H oxidase activity and to intrinsic antioxidant properties of the drugs.

CHAPTER 13

Assessing Reactive Oxygen Species in the Cardiovascular System

To better understand the biology of ROS and the pathophysiological implications in cardiovascular disease, it is essential to accurately measure ROS levels and to carefully characterize the sources from which ROS are derived. Many assays and strategies are currently available, however, they all have limitations and the ideal methods have yet to be established. The most commonly used methods in biological systems are those based on spectrophotometry, chemiluminesence, electron spin resonance and fluorescence. Factors that dictate which method to use include: the type of species to be measured, the preparation in which ROS are measured (tissue, cells, membranes, organelles, whole animals, humans), fresh versus frozen samples and intracellular versus extracellular ROS. An ideal assay should be sensitive enough to measure ROS in the linear range of that assay and the assay should be robust and reproducible [386–388]. It should be specific for the type of ROS to be measured, and the signal should be inhibited by a specific scavenger for that particular species. Because of the complexity relating to ROS measurement, it is suggested that a combination of techniques are used, with at least two different assays being employed [386, 388].

Specific challenges that warrant consideration when measuring ROS in cardiovascular tissue include: (1) the tissue/cells/system that is studied may not produce ROS at levels that are detectable; (2) ROS are highly compartmentalized and highly localized; (3) addition of scavengers (whether they are capable of crossing the membrane or not) might not block the effect that is being assessed; (4) some of the scavengers that are available may have their effect by removing one ROS but producing another with similar biological effects (e.g., SOD); (5) ROS are unstable and highly reactive with a short half life; (6) ROS can also be formed at high levels at the site of production (cells, organelles), but this may not be reflected in the intact sample (tissue, whole cell, or media); and (7) transition metals that are present in the buffers used for ROS assays can produce an artefact. Another issue arises when attempting to measure ROS in intact tissue. For some assays, such as lucigenin-enhanced chemiluminescence, it is necessary to add the enzyme substrate (e.g., NADPH), which may not be cell membrane permeable and hence the substrate does not access intracellular oxidases. It is also important to note that most of the techniques currently available are either qualitative or semi-quantitative at best, and some of the probes used for these assays are non-specific [386, 387].

Some commonly used assays to measure ROS concentrations in cardiovascular tissue are summarized below.

13.1 CYTOCHROME C REDUCTION

Oxygen possesses two unpaired electrons, and when it gains an additional electron, $\bullet O_2^-$ is formed. Superoxide donates its extra electron to other molecules, reducing it and the product for this reaction, in some cases, can be specifically measured by spectrophotometry [388, 389]. Ferricytochrome C can be reduced by superoxide to ferrocytochrome C, and when this happens, the spectrophotometric absorbance of ferrocytochrome C at 550 nm is increased, whereas at 540 and 560 nm remain unchanged. This is a very good technique for quantification of $\bullet O_2^-$, however, because ferricytochrome C can be easily reduced by other enzymes and molecules, it is imperative that the protocol is performed in the presence and absence of SOD, where only the signal inhibited by SOD should be used for final calculations of $\bullet O_2^-$ concentrations. This technique allows the quantification of $\bullet O_2^-$ within the picomolar ranges, it is relatively easy and does not require specialized equipment. To prevent reoxidation of ferricytochrome C by H_2O_2, it is suggested that experiments are performed in the presence of catalase.

13.2 LUCIGENIN-ENHANCED CHEMILUMINESCENCE

Chemiluminescence protocols are based on the property that chemiluminescence probes release photons after exposure to $\bullet O_2^-$. The lucigenin-enhanced chemiluminescence technique is the most frequently used method in assessing NADPH oxidase-generated ROS in cardiovascular tissue, where superoxide reacts with lucigenin leading to the formation of a reduced cation radical of lucigenin [390, 391]. The product of this first reaction reacts again with another $\bullet O_2^-$, forming the dioxetane molecule, which emits a photon that can be detected by a luminometer or scintillation counter. Usually, chemiluminescence probes are cell permeable, which means that the levels of $\bullet O_2^-$ measured are a reflection of its production in the extracellular and intracellular milieu.

Studies, in artificial cell systems, demonstrated that the lucigenin radical can react with O_2 to form $\bullet O_2^-$, overestimating the real levels of this ROS in the assay. This reaction is known as the redox cycling of lucigenin and is catalyzed by flavin containing enzymes (nitric oxide synthase, xanthine oxidase and cytochrome P450 monooxygenases). However, when lucigenin is used at low concentrations (5 μM), the artefact induced by the redox cycling of lucigenin is insignificant. Accordingly, experiments should be performed using lucigenin at concentrations ≤ 5 μM. Other luminescence probes that have been used to assess ROS include luminol (5-amino-2,3-dihydro-1,

4-phthalazinedione) coelenterazine and MCLA (methyladed cypridina luciferin analog) [392]. However, these probes are less specific for $\bullet O_2^-$, and they are prone to redox cycling.

13.3   DIHYDROETHIDIUM AND HIGH-PERFORMANCE LIQUID CHROMATOGRAPHY

Dihydroethidium (DHE), a cell permeable molecule, reacts with $\bullet O_2^-$ and once oxidized forms ethidium bromide, a fluophore that binds to DNA [393]. This reaction was considered fairly specific for $\bullet O_2^-$ but it has been shown that other ROS can also oxidize DHE, such as H_2O_2 and ONOO−. Since DHE can cross the plasma membrane, the measurement of DHE fluorescence is an index of intracellular $\bullet O_2^-$ generation and can be assessed by flow cytometry or by fluorescence microscopy. It is important to note that DHE does not undergo redox cycling. It has become increasingly apparent that DHE reaction with $\bullet O_2^-$ to form ethidium bromide is not specific. The product from the oxidation of DHE by $\bullet O_2^-$ is 2-hydroxyethidium and not ethidium bromide. 2-hydroxyethidium formation is very specific to $\bullet O_2^-$ and has a molecular weight higher than DHE, whereas ethidium production seems to reflect the redox status of the tissue or cell. Nowadays, using high-performance liquid chromatography (HPLC), DHE (non-fluorescent), 2-hydroxyethidium (highly fluorescent) and ethidium (fluorescent) can be separated and the 2-hydroxyethidium peak can be analyzed to assess the $\bullet O_2^-$ concentration accurately [394, 395].

13.4   DICHLOROFLUORESCEIN AND AMPLEX RED TO MEARURE H_2O_2

As for $\bullet O_2^-$, there are fluorescence-based probes to measure H_2O_2. 2'7'-dichlorofluorescein di-acetate (DCFH-DA) is a probe that once crosses the plasma membrane is cleaved by esterases to DCFH and then is trapped within the cell [396]. DCFH is then oxidized by H_2O_2 to DCF, which is highly fluorescent. However, H_2O_2 may not be the only ROS to oxidize DCFH. DCFH-DA also reacts with ONOO−, lipid peroxidises and to a lesser extent $\bullet O_2^-$. These considerations need to be kept in mind when using DCFH-DA as a tool to measure H_2O_2. This is a good technique to evaluate intracellular ROS and its subcellular localization since DCFH is plasma membrane impermeable.

Amplex Red, an easy and straightforward assay developed by Molecular Probes, was released and is considered by many researchers a useful assay for H_2O_2 that is produced and released by cells and tissues. Hydrogen peroxide levels are assessed by horseradish peroxidase-catalyzed oxidation of a colorless and non-fluorescent molecule N-acetyl-3,7-dihydroxyphenoxazine (Amplex Red) to resorufin, which is very stable. The molecule resorufin emits light when it is excited at 530 nm. This assay seems to be specific and sensitive for H_2O_2, with stoichiometry of 1:1. One advantage is the fact that this assay is commercially available, but Amplex Red dye is air- and light-sensitive, which makes it unstable, and hence the reagent needs to be used immediately upon opening the sealed product. To avoid artefacts, one should not use high concentrations (>50 μM) of Amplex Red because this dye can be auto-oxidized and generate $\bullet O_2^-$ and H_2O_2.

13.5 DIHYDRORHODAMINE OXIDATION TO MEASURE ONOO–

Dihydrorhodamine 123 (DHR) is an analog of DCFH-DA that can undergo oxidation to rhodamine 123, which is a fluophore and its signal can be measured in a fluorescence microscope or fluorimeter [397]. Rhodamine 123 is retained in the intracellular space. DHR 123 is oxidized by peroxynitrite, but as with other probes can also be oxidized by other ROS, such as OH$\bullet$ and H_2O_2. Accordingly, the use of inhibitors, such as SOD mimetics, catalase, nitric oxide inhibitors and peroxynitrite scavengers, is imperative.

13.6 ELECTRON SPIN RESONANCE

Electron paramagnetic resonance (EPR) spectroscopy (also known as electron spin resonance (ESR)) provides the most direct insight into the biochemistry of free radicals and antioxidants [398]. The technique requires specialized equipment and expertise and hence is not used routinely in cardiovascular biology studies. However, this method is considered the gold standard in redox research, and the reader is referred to many excellent publications in the literature for further details [399–402].

Each assay has strengths and limitations, and the appropriate assays for use need to be selected carefully. The reader is referred to some excellent reviews on different approaches to measure ROS in the cardiovascular system [386, 387, 403].

· · · ·

CHAPTER 14

Assessing ROS in Clinical Studies

Investigating oxidative stress in relation to human disease is even more challenging than that in experimental systems, because many of the currently available methods, as noted above, are not generally applicable to clinical examination. In the clinic, ROS status is usually assessed in plasma or urine by measuring the levels of oxidation target product (e.g., Maolondialdehyde (MDA), 3-nitrotyrosine, protein carbonyls) or anti-oxidants (e.g., thiols, vitamins, bilirubin, dlutathione, catalase, SOD) as biomarkers of oxidative status [404]. In general, an increase in ROS bioavailability is reflected as an increase in biomarker levels of oxidation and a decrease in anti-oxidant levels. Biomarkers of oxidative stress used in clinical studies are detailed in Table 1. It should be

TABLE 1: Biomarkers of oxidative stress used in clinical studies.
PROTEIN OXIDATION
Oxidized tyrosine, methionine, tryptophan
Nitrate proteins
Protein carbonyls
SH-oxidation
LIPID OXIDATION
Malondialdehyde (MDA) (thiobarbituric acid-reactive substances (TBARS))
Isoprostanes
DNA OXIDATION
Oxidized bases
Nitrated bases

TABLE 2: Characteristics for a reliable biomarker of oxidative stress as described by Giustarini et al. [254].
1. Stable molecule
2. Implicated directly in the disease process
3. Specific for the ROS/RNS under investigation
4. Non-invasive measurement
5. Low intra- and inter-variability
6. Accurate, precise, specific, sensitive, validated assays
7. Quantitative rather than qualitative assay
8. Consensus and establishment of reference values and ranges

stressed that there are many analytical and methodological limitations in measuring ROS clinically, and there is still no consensus as to which biomarkers are the most appropriate in assessing redox status in humans. In fact, the "ideal biomarker" as described by Giustarini et al. [254] has yet to be determined (Table 2). The reader is referred to an excellent review by this group, which critically discusses the challenges regarding ROS measurements in the clinic [254]. In general, it is suggested that a panel of biomarkers in multiple systems (plasma, urine, tissue, cells) be used.

·   ·   ·   ·

CHAPTER 15

Conclusion

In physiological conditions, ROS play an important role in cardiovascular and renal biology and in autonomic control. In the vasculature, ROS regulate endothelial function and vascular tone through highly controlled redox-sensitive signalling pathways. Uncontrolled production/degradation of ROS results in oxidative stress, which induces cardiovascular and renal damage and activation of the sympathetic nervous system with associated increase in blood pressure. Although oxidative damage may not be the sole cause of blood pressure elevation, together with pro-hypertensive factors, such as salt-loading, activation of the renin–angiotensin system and sympathetic hyperactivity, it augments the development of hypertension. Convincing findings from experimental and animal studies suggest a causative role for oxidative stress in the pathogenesis of hypertension. However, in humans, there is still no solid evidence that oxidative stress is fundamentally involved in the pathogenesis of hypertension. Further research in the field of oxidative stress and human hypertension is warranted. In particular, there is a need for the development of sensitive, specific and reliable biomarkers and assays to assess the redox status of patients.

· · · ·

Acknowledgments

Work from the author's laboratory was supported by grants from the Canadian Institutes of Health Research, Heart and Stroke Foundation of Canada, Kidney Foundation of Canada/Pfizer and the Juvenile Diabetes Research Foundation.

Conflicts: There are no conflicts to declare

References

[1] Greabu, M., Battino, M., Mohora, M., Olinescu, R., Totan, A., Didilescu, A. Oxygen, a paradoxical element? *Rom J Intern Med* 46(2): pp. 125–35, 2008.

[2] Touyz, R. M., Schiffrin, E. L. Reactive oxygen species in vascular biology: implications in hypertension. *Histochem Cell Biol* 122(4): pp. 339–52, 2004.

[3] Droge, W. Free radicals in the physiological control of cell function. *Physiol Rev* 82(1): pp. 47–95, 2002.

[4] Mueller, C. F., Laude, K., McNally, J. S., Harrison, D. G. ATVB in focus: redox mechanisms in blood vessels. *Arterioscler Thromb Vasc Biol* 25(2): pp. 274–8, 2005.

[5] Pawlak, K., Naumnik, B., Brzosko, S., Pawlak, D., Mysliwiec, M. Oxidative stress—a link between endothelial injury, coagulation activation, and atherosclerosis in haemodialysis patients. *Am J Nephrol* 24(1): pp. 154–61, 2004.

[6] Tain, Y. L., Baylis, C. Dissecting the causes of oxidative stress in an in vivo model of hypertension. *Hypertension* 48(5): pp. 828–9, 2006.

[7] Vaziri, N. D., Rodriguez-Iturbe, B. Mechanisms of disease: oxidative stress and inflammation in the pathogenesis of hypertension. *Nat Clin Pract Nephrol* 2(10): pp. 582–91, 2006.

[8] Landmesser, U., Harrison, D. G., Drexler, H. Oxidant stress-a major cause of reduced endothelial nitric oxide availability in cardiovascular disease. *Eur J Clin Pharmacol* pp. 62:13–9, 2006.

[9] Ushio-Fukai, M., Alexander, R. W., Akers, M., Griendling, K. K. p38 Mitogen-activated protein kinase is a critical component of the redox-sensitive signaling pathways activated by angiotensin II. Role in vascular smooth muscle cell hypertrophy. *J Biol Chem* 273: pp. 15022–9, 1998.

[10] Griendling, K. K., Sorescu, D., Lassegue, B., Ushio-Fukai, M. Modulation of protein kinase activity and gene expression by reactive oxygen species and their role in vascular physiology and pathophysiology. *Arterioscler Thromb Vasc Biol* 20: pp. 2175–83, 2000.

[11] Zhang, Y., Griendling, K. K., Dikalova, A., Owens, G. K., Taylor, W. R. Vascular hypertrophy in angiotensin II-induced hypertension is mediated by vascular smooth muscle cell-derived H2O2. *Hypertension* 46: pp. 732–7, 2005.

[12]   Hool, L. C., Corry, B. Redox control of calcium channels: from mechanisms to therapeutic opportunities. *Antioxid Redox Signal* 9(4): pp. 409–35, 2007.

[13]   Touyz, R. M., Tabet, F., Schiffrin, E. L. Redox-dependent signalling by angiotensin II and vascular remodelling in hypertension. *Clin Exp Pharmacol Physiol* 30(11): pp. 860–6, 2003.

[14]   Touyz, R. M. Reactive oxygen species as mediators of calcium signalling by angiotensin II: implications in vascular physiology and pathophysiology. *Antioxid Redox Signal* 7(9–10): pp. 1302–14, 2005.

[15]   Millar, T. M., Phan, V., Tibbles, L. A. ROS generation in endothelial hypoxia and reoxygenation stimulates MAP kinase signaling and kinase-dependent neutrophil recruitment. *Free Radic Biol Med* 42(8): pp. 1165–677, 2007.

[16]   Kimura, S., Zhang, G. X., Nishiyama, A., Shokoji, T., Yao, L., Fan, Y. Y., Rahman, M., Abe, Y. Mitochondria-derived reactive oxygen species and vascular MAP kinases: comparison of angiotensin II and diazoxide. *Hypertension* 45(3): pp. 438–44, 2005.

[17]   Tabet, F., Savoia, C., Schiffrin, E. L., Touyz, R. M. Differential calcium regulation by hydrogen peroxide and superoxide in vascular smooth muscle cells from spontaneously hypertensive rats. *J Cardiovasc Pharmacol* 44(2): pp. 200–8, 2004.

[18]   Gutierrez, J., Ballinger, S. W., Darley-Usmar, V. M., Landar, A. Free radicals, mitochondria, and oxidized lipids: the emerging role in signal transduction in vascular cells. *Circ Res* 99(9): pp. 924–32, 2006.

[19]   Usatyuk, P. V., Parinandi, N. L., Natarajan, V. Redox regulation of 4-hydroxy-2-nonenal-mediated endothelial barrier dysfunction by focal adhesion, adherens, and tight junction proteins. *J Biol Chem* 281(46): pp. 35554–66, 2006.

[20]   Yoshioka, J., Schreiter, E. R., Lee, R. T. Role of thioredoxin in cell growth through interactions with signaling molecules. *Antioxid Redox Signal* 8(11–12): pp. 2143–51, 2006.

[21]   Anathy, V., Aesif, S. W., Guala, A. S., Havermans, M., Reynaert, N. L., Ho, Y. S., Budd, R. C., Janssen-Heininger, Y. M. Redox amplification of apoptosis by caspase-dependent cleavage of glutaredoxin 1 and S-glutathionylation of Fas. *J Cell Biol* 184(2): pp. 241–52, 2009.

[22]   Romanowski, A., Murray, I. R., Huston, M. J. Effects of hydrogen peroxide on normal and hypertensive rats. *Pharm Acta Helv* 35: pp. 354–7, 1960.

[23]   Rajagopalan, S., Kurz, S., Munzel, T., Tarpey, M., Freeman, B. A., Griendling, K. K. Angiotensin II-mediated hypertension in the rat increases vascular superoxide production via membrane NADH/NAD(P)H oxidase activation. Contribution to alterations of vasomotor tone. *J Clin Invest* 97: pp. 1916–23, 1996.

[24]   Zalba, G., Beaumont, F. J., San Jose, G., Fortuno, A., Fortuno, M. A., Etayo, J. C., et al. Vascular NADH/NAD(P)H oxidase is involved in enhanced superoxide production in spontaneously hypertensive rats. *Hypertension* 35: pp. 1055–61, 2000.

[25]   Akasaki, T., Ohya, Y., Kuroda, J., Eto, K., Abe, I., Sumimoto, H., Iida, M. Increased expression of gp91phox homologues of NAD(P)H oxidase in the aortic media during chronic hypertension: involvement of the renin–angiotensin system. *Hypertens Res* 29(10): pp. 813–20, 2006.

[26]   Fujita, M., Ando, K., Kawarazaki, H., Kawarasaki, C., Muraoka, K., Ohtsu, H., Shimizu, H., Fujita, T. Sympathoexcitation by brain oxidative stress mediates arterial pressure elevation in salt-induced chronic kidney disease. *Hypertension* 59(1): pp. 105–12, 2012.

[27]   Kagota, S., Tada, Y., Kubota, Y., Nejime, N., Yamaguchi, Y., Nakamura, K., Kunitomo, M., Shinozuka, K. Peroxynitrite is Involved in the dysfunction of vasorelaxation in SHR/NDmcr-cp rats, spontaneously hypertensive obese rats. *J Cardiovasc Pharmacol* 50(6): pp. 677–85, 2007.

[28]   Klanke, B., Cordasic, N., Hartner, A., Schmieder, R. E., Veelken, R., Hilgers, K. F. Blood pressure versus direct mineralocorticoid effects on kidney inflammation and fibrosis in DOCA-salt hypertension. *Nephrol Dial Transplant* 23(11): pp. 3456–63, 2008.

[29]   Landmesser, U., Cai, H., Dikalov, S., McCann, L., Hwang, J., Jo, H., et al. Role of p47(phox) in vascular oxidative stress and hypertension caused by angiotensin II. *Hypertension* 40: pp. 511–5, 2002.

[30]   Jung, O., Schreiber, J. G., Geiger, H., Pedrazzini, T., Busse, R., Brandes, R. P. gp91phox-containing NAD(P)H oxidase mediates endothelial dysfunction in renovascular hypertension. *Circulation* 109(14): pp. 1795–801, 2004.

[31]   Lavi, S., Yang, E. H., Prasad, A., Mathew, V., Barsness, G. W., Rihal, C. S., Lerman, L. O., Lerman, A. The interaction between coronary endothelial dysfunction, local oxidative stress, and endogenous nitric oxide in humans. *Hypertension* 51(1): pp. 127–33, 2008.

[32]   Franco, M. C., Kawamoto, E. M., Gorjão, R., Rastelli, V. M., Curi, R., Scavone, C., Sawaya, A. L., Fortes, Z. B., Sesso, R. Biomarkers of oxidative stress and antioxidant status in children born small for gestational age: evidence of lipid peroxidation. *Pediatr Res* 62(2) pp. 204–8, 2007.

[33]   Cottone, S., Mulè G, Guarneri, M., Palermo, A., Lorito, M. C., Riccobene, R., Arsena, R., Vaccaro, F., Vadalà, A., Nardi, E., Cusimano, P, Cerasola, G. Endothelin-1 and F2-isoprostane relate to and predict renal dysfunction in hypertensive patients. *Nephrol Dial Transplant* 24(2): pp. 497–503, 2009.

[34]   Mistry, H. D., Wilson, V., Ramsay, M. M., Symonds, M. E., Broughton Pipkin, F. Reduced selenium concentrations and glutathione peroxidase activity in preeclamptic pregnancies. *Hypertension* 52(5): pp. 881–8, 2008.

[35]   Duffy, S. J., Gokce, N., Holbrook, M., Huang, A., Frei, B., Keaney, J. F. Jr, Vita, J. A. Treatment of hypertension with ascorbic acid. *Lancet* 354(9195): pp. 2048–9, 1999.

[36]   Duffy, S. J., Gokce, N., Holbrook, M., Hunter, L. M., Biegelsen, E. S., Huang, A., Keaney, J. F. Jr, Vita, J. A. Effect of ascorbic acid treatment on conduit vessel endothelial dysfunction in patients with hypertension. *Am J Physiol Heart Circ Physiol* 280(2): pp. H528–34, 2001.

[37]   Kurl, S., Tuomainen, T. P., Laukkanen, J. A., Nyyssönen, K., Lakka, T., Sivenius, J., Salonen, J. T. Plasma vitamin C modifies the association between hypertension and risk of stroke. *Stroke* 33(6): pp. 1568–73, 2002.

[38]   Hajjar, I. M., George, V., Sasse, E. A., Kochar, M. S. A randomized, double-blind, controlled trial of vitamin C in the management of hypertension and lipids. *Am J Ther* 9(4): pp. 289–93, 2002.

[39]   Svetkey, L. P., Loria, C. M. Blood pressure effects of vitamin C: what's the key question? *Hypertension* 40(6): pp. 789–91, 2002.

[40]   Darko, D., Dornhorst, A., Kelly, F. J., Ritter, J. M., Chowienczyk, P. J. Lack of effect of oral vitamin C on blood pressure, oxidative stress and endothelial function in Type II diabetes. *Clin Sci (Lond)* 103(4): pp. 339–44, 2002.

[41]   Hatzitolios, A., Iliadis, F., Katsiki, N., Baltatzi, M. Is the anti-hypertensive effect of dietary supplements via aldehydes reduction evidence based? A systematic review. *Clin Exp Hypertens* 30(7): pp. 628–39, 2008.

[42]   Wray, D. W., Uberoi, A., Lawrenson, L., Bailey, D. M., Richardson, R. S. Oral antioxidants and cardiovascular health in the exercise-trained and untrained elderly: a radically different outcome. *Clin Sci (Lond)* 116(5): pp. 433–41, 2009.

[43]   Fridovich, I. Superoxide anion radical (O2-.), superoxide dismutases, and related matters. *J Biol Chem* 272(30): pp. 18515–7, 1997.

[44]   Johnson, F., Giulivi, C. Superoxide dismutases and their impact upon human health. *Mol Aspects Med* 26(4–5): pp. 340–52, 2005.

[45]   Forman, H. J., Fukuto, J. M., Miller, T., Zhang, H., Rinna, A., Levy, S. The chemistry of cell signaling by reactive oxygen and nitrogen species and 4-hydroxynonenal. *Arch Biochem Biophys* 477(2): pp. 183–95, 2008.

[46]   Jacob, C., Jamier, V., Ba, L. A. Redox active secondary metabolites. *Curr Opin Chem Biol* 15(1): pp. 149–55, 2011.

[47]   Raha, S., Myint, A. T., Johnstone, L., Robinson, B. H. Control of oxygen free radical formation from mitochondrial complex I: roles for protein kinase A and pyruvate dehydrogenase kinase. *Free Radic Biol Med* 32(5): pp. 421–30, 2002.

[48]   Babior, B. M. NADPH oxidase. *Curr Opin Immunol* 16(1): pp. 42–7, 2004.

[49]   Brown, D. I., Griendling, K. K. Nox proteins in signal transduction. *Free Radic Biol Med* 47(9): pp. 1239–53, 2009.

[50]   Ferrer-Sueta, G., Radi, R. Chemical biology of peroxynitrite: kinetics, diffusion, and radicals. *ACS Chem Biol* 4(3): pp. 161–77, 2009.

[51] van der Donk, W. A., Krebs, C., Bollinger, J. M. Jr. Substrate activation by iron superoxo intermediates. *Curr Opin Struct Biol* 20(6): pp. 673–83, 2010.

[52] Lindahl, M., Mata-Cabana, A., Kieselbach, T. The disulfide proteome and other reactive cysteine proteomes: analysis and functional significance. *Antioxid Redox Signal* 14(12): pp. 2581–642, 2011.

[53] Faraci, F. M., Didion, S. P. Vascular protection: superoxide dismutase isoforms in the vessel wall. *Arterioscler Thromb Vasc Biol* 24(8): pp. 1367–73, 2004.

[54] Mendez, J. I., Nicholson, W. J., Taylor, W. R. SOD isoforms and signaling in blood vessels: evidence for the importance of ROS compartmentalization. *Arterioscler Thromb Vasc Biol* 25(5): pp. 887–8, 2005.

[55] Welch, W. J., Chabrashvili, T., Solis, G., Chen, Y., Gill, P. S., Aslam, S., Wang, X., Ji, H., Sandberg, K., Jose, P., Wilcox, C. S. Role of extracellular superoxide dismutase in the mouse angiotensin slow pressor response. *Hypertension* 48(5): pp. 934–41, 2006 Nov.

[56] Lob, H. E., Vinh, A., Li, L., Blinder, Y., Offermanns, S., Harrison, D. G. Role of vascular extracellular superoxide dismutase in hypertension. *Hypertension* 58(2): pp. 232–9, 2011.

[57] Hawkins, B. J., Madesh, M., Kirkpatrick, C. J., Fisher, A. B. Superoxide flux in endothelial cells via the chloride channel-3 mediates intracellular signaling. *Mol Biol Cell* 18(6): pp. 2002–12, 2007.

[58] Miller, F. J. Jr, Filali, M., Huss, G. J., Stanic, B., Chamseddine, A., Barna, T. J., Lamb, F. S. Cytokine activation of nuclear factor kappa B in vascular smooth muscle cells requires signaling endosomes containing Nox1 and ClC-3. *Circ Res* 101(7): pp. 663–71, 2007.

[59] Chu, X., Filali, M., Stanic, B., Takapoo, M., Sheehan, A., Bhalla, R., Lamb, F. S., Miller, F. J. Jr. A critical role for chloride channel-3 (CIC-3) in smooth muscle cell activation and neointima formation. *Arterioscler Thromb Vasc Biol* 31(2): pp. 345–51, 2011.

[60] Mittler, R., Vanderauwera, S., Suzuki, N., Miller, G., Tognetti, V. B., Vandepoele, K., Gollery, M., Shulaev, V., Van Breusegem, F. ROS signaling: the new wave? *Trends Plant Sci* 16(6): pp. 300–9, 2011.

[61] Finkel, T. Signal transduction by reactive oxygen species. *J Cell Biol* 194(1): pp. 7–15, 2011.

[62] Zinkevich, N. S., Gutterman, D. D. ROS-induced ROS release in vascular biology: redox-redox signaling. *Am J Physiol Heart Circ Physiol* 301(3): pp. H647–53, 2011.

[63] Cai, H., Harrison, D. G. Endothelial dysfunction in cardiovascular diseases: the role of oxidant stress. *Circ Res* 87: pp. 840–4, 2000.

[64] Cai, H. Hydrogen peroxide regulation of endothelial function: origins, mechanisms, and consequences. *Cardiovasc Res* 68(1): pp. 26–36, 2005.

[65] Paravicini, T. M., Chrissobolis, S., Drummond, G. R., Sobey, C. G. Increased NAD(P)H-oxidase activity and Nox4 expression during chronic hypertension is associated with enhanced cerebral vasodilatation to NAD(P)H in vivo. *Stroke* 35: pp. 584–9, 2004.

[66]   Liu, H., Li, H., Kalyanaraman, B., Nicolosi, A. C., Gutterman, D. D. Mitochondrial sources of H_2O_2 generation play a key role in flow-mediated dilation in human coronary resistance arteries. *Circ Res* 93: pp. 573–80, 2003.

[67]   Matoba, T., Shimokawa, H., Nakashima, M., Hirakawa, Y., Mukai, Y., Hirano, K., et al. Hydrogen peroxide is an endothelium-derived hyperpolarizing factor in mice. *J Clin Invest* 106: pp. 1521–30, 2000.

[68]   Ignarro, L. J. Nitric oxide: a unique endogenous signaling molecule in vascular biology. *Biosci Rep* 19(2): pp. 51–71, 1999.

[69]   Napoli, C., Ignarro, L. J. Nitric oxide and pathogenic mechanisms involved in the development of vascular diseases. *Arch Pharm Res* 32(8): pp. 1103–8, 2009.

[70]   Culotta, E., Koshland, D. E. Jr. NO news is good news. Science. 258(5090): pp. 1862–5, 1992.

[71]   Villanueva, C., Giulivi, C. Subcellular and cellular locations of nitric oxide synthase isoforms as determinants of health and disease. *Free Radic Biol Med* 49(3): pp. 307–16, 2010.

[72]   Maia, L. B., Moura, J. J. Nitrite reduction by xanthine oxidase family enzymes: a new class of nitrite reductases. *J Biol Inorg Chem* 16(3): pp. 443–60, 2011.

[73]   Allen, B. W., Demchenko, I. T., Piantadosi, C. A. Two faces of nitric oxide: implications for cellular mechanisms of oxygen toxicity. *J Appl Physiol* 106(2): pp. 662-, 2009.

[74]   Szabó C, Ischiropoulos, H., Radi, R. Peroxynitrite: biochemistry, pathophysiology and development of therapeutics. *Nat Rev Drug Discov* 6(8): pp. 662–80, 2007.

[75]   Zhang, Y., Janssens, S. P., Wingler, K., Schmidt, H. H., Moens, A.L. Modulating endothelial nitric oxide synthase: a new cardiovascular therapeutic strategy. *Am J Physiol Heart Circ Physiol* 301(3): pp. H634–46, 2011.

[76]   Nishino, T., Okamoto, K., Eger, B. T., Pai, E. F., Nishino, T. Mammalian xanthine oxidoreductase—mechanism of transition from xanthine dehydrogenase to xanthine oxidase. *FEBS J* 275(13): pp. 3278–89, 2008.

[77]   Doehner W., Landmesser U. Xanthine oxidase and uric acid in cardiovascular disease: clinical impact and therapeutic options. *Semin Nephrol.* 2011;31(5): pp. 433–40.

[78]   Choi H., Tostes R. C., Webb R. C. Mitochondrial aldehyde dehydrogenase prevents ROS-induced vascular contraction in angiotensin-II hypertensive mice. *J Am Soc Hypertens.* 2011;5(3): pp. 154–60

[79]   Moens, A. L., Kass, D. A. Tetrahydrobiopterin and cardiovascular disease. *Arterioscler Thromb Vasc Biol* 26(11): pp. 2439–44, 2006.

[80]   Mukhopadhyay, P., Rajesh, M., Bátkai, S., Kashiwaya, Y., Haskó G., Liaudet, L., Szabó C., Pacher, P. Role of superoxide, nitric oxide, and peroxynitrite in doxorubicin-induced cell death in vivo and in vitro. *Am J Physiol Heart Circ Physiol* 296(5): pp. H1466–83, 2009 May.

[81] Förstermann, U. Oxidative stress in vascular disease: causes, defense mechanisms and potential therapies. *Nat Clin Pract Cardiovasc Med* 5(6): pp. 338–49, 2008.

[82] Suzuki, H., DeLano, F. A., Parks, D. A., Jamshidi, N., Granger, D. N., Ishii, H., et al. Xanthine oxidase activity associated with arterial blood pressure in spontaneously hypertensive rats. *Proc Natl Acad Sci U S A* 95: pp. 4754–9, 1998.

[83] DeLano, F. A., Parks, D. A., Ruedi, J. M., Babior, B. M., Schmid-Schonbein, G.W. Microvascular display of xanthine oxidase and NAD(P)H oxidase in the spontaneously hypertensive rat. *Microcirculation* 13(7): pp. 551–66, 2006.

[84] Mervaala, E. M., Cheng, Z. J., Tikkanen, I., Lapatto, R., Nurminen, K., Vapaatalo, H. Endothelial dysfunction and xanthine oxidoreductase activity in rats with human renin and angiotensinogen genes. *Hypertension* 37: pp. 414–8, 2001.

[85] Laakso, J., Mervaala, E., Himberg, J. J., Teravainen, T. L., Karppanen, H., Vapaatalo, H., et al. Increased kidney xanthine oxidoreductase activity in salt-induced experimental hypertension. *Hypertension* 32: pp. 902–6, 1998.

[86] Siu, Y. P., Leung, K. T., Tong, M. K., Kwan, T. H. Use of allopurinol in slowing the progression of renal disease through its ability to lower serum uric acid level. *Am J Kidney Dis* 47(1): pp. 51–9, 2006.

[87] Laakso, J. T., Teravainen, T. L., Martelin, E., Vaskonen, T., Lapatto, R. Renal xanthine oxidoreductase activity during development of hypertension in spontaneously hypertensive rats. *J Hypertens* 22: pp. 1333–40, 2004.

[88] Corry, D. B., Tuck, M. L. Uric acid and the vasculature. *Curr Hypertens Rep* 8(2): pp. 116–9, 2006.

[89] Feig, D. I., Soletsky, B., Johnson, R. J. Effect of allopurinol on blood pressure of adolescents with newly diagnosed essential hypertension: a randomized trial. *JAMA* 300(8): pp. 924–32, 2008.

[90] Andrew, P. J., Mayer, B. Enzymatic function of nitric oxide synthases. *Cardiovasc Res* 43: pp. 521–31, 1999.

[91] Vasquez-Vivar, J., Kalyanaraman, B., Martasek, P., Hogg, N., Masters, B. S., Karoui, H., et al. Superoxide generation by endothelial nitric oxide synthase: the influence of cofactors. *Proc Natl Acad Sci U S A* 95: pp. 9220–5, 1998.

[92] Landmesser, U., Dikalov, S., Price, S. R., McCann, L., Fukai, T., Holland, S. M., et al. Oxidation of tetrahydrobiopterin leads to uncoupling of endothelial cell nitric oxide synthase in hypertension. *J Clin Invest* 111: pp. 1201–9, 2003.

[93] Hong, H. J., Hsiao, G., Cheng, T. H., Yen, M. H. Supplemention with tetrahydrobiopterin suppresses the development of hypertension in spontaneously hypertensive rats. *Hypertension* 38: pp. 1044–8, 2001.

[94]   Ketonen, J., Mervaala, E. Effects of dietary sodium on reactive oxygen species formation and endothelial dysfunction in low-density lipoprotein receptor-deficient mice on high-fat diet. *Heart Vessels* 23(6): pp. 420–9, 2008.

[95]   Moens, A. L., Takimoto, E., Tocchetti, C. G., Chakir, K., Bedja, D., Cormaci, G., Ketner, E. A., Majmudar, M., Gabrielson, K., Halushka, M. K., Mitchell, J. B., Biswal, S., Channon, K. M., Wolin, M. S., Alp, N. J., Paolocci, N., Champion, H. C., Kass, D. A. Reversal of cardiac hypertrophy and fibrosis from pressure overload by tetrahydrobiopterin: efficacy of recoupling nitric oxide synthase as a therapeutic strategy. *Circulation* 117(20): pp. 2626–36, 2008.

[96]   Moens, A. L., Kass, D. A. Therapeutic potential of tetrahydrobiopterin for treating vascular and cardiac disease. *J Cardiovasc Pharmacol* 50(3): pp. 238–46, 2007.

[97]   Adlam, D., Bendall, J. K., De Bono, J. P., Alp, N. J., Khoo, J., Nicoli, T., Yokoyama, M., Kawashima, S., Channon, K. M. Cardiovascular control: relationships between nitric oxide-mediated endothelial function, eNOS coupling and blood pressure revealed by eNOS-GTP cyclohydrolase 1 double transgenic mice. *Exp Physiol* 92(1): pp. 119–26, 2007.

[98]   Bauersachs, J., Widder, J. D. Tetrahydrobiopterin, endothelial nitric oxide synthase, and mitochondrial function in the heart. *Hypertension* 53(6): pp. 907–8, 2009.

[99]   Katusic, Z. S., d'Uscio, L. V., Nath, K. A. Vascular protection by tetrahydrobiopterin: progress and therapeutic prospects. *Trends Pharmacol Sci* 30(1): pp. 48–54, 2009.

[100]   Wang, S., Xu, J., Song, P., Wu, Y., Zhang, J., Chul Choi, H., Zou, M. H. Acute inhibition of guanosine triphosphate cyclohydrolase 1 uncouples endothelial nitric oxide synthase and elevates blood pressure. *Hypertension* 52(3): pp. 484–90, 2008.

[101]   Porkert, M., Sher, S., Reddy, U., Cheema, F., Niessner, C., Kolm, P., Jones, D. P., Hooper, C., Taylor, W. R., Harrison, D., Quyyumi, A. A. Tetrahydrobiopterin: a novel antihypertensive therapy. *J Hum Hypertens* 22(6): pp. 401–7, 2008.

[102]   Oelze, M., Daiber, A., Brandes, R. P., Hortmann, M., Wenzel, P., Hink, U., Schulz, E., Mollnau, H., von Sandersleben, A., Kleschyov, A. L., Mülsch, A., Li, H., Förstermann, U., Münzel, T. Nebivolol inhibits superoxide formation by NAD(P)H oxidase and endothelial dysfunction in angiotensin II-treated rats. *Hypertension* 48(4): pp. 677–84, 2006.

[103]   de Cavanagh, E. M., Inserra, F., Ferder, M., Ferder, L. From mitochondria to disease: role of the renin–angiotensin system. *Am J Nephrol* 27(6): pp. 545–53, 2007.

[104]   Duszynski, J., Koziel, R., Brutkowski, W., Szczepanowska, J., Zablocki, K. The regulatory role of mitochondria in capacitative calcium entry. *Biochim Biophys Acta* 1757(5–6): pp. 380–7, 2006.

[105]   Thomas, D. D., Liu, X., Kantrow, S. P., Lancaster, J. R. Jr. The biological lifetime of nitric oxide: implications for the perivascular dynamics of NO and O2. *Proc Natl Acad Sci U S A* 98(1): pp. 355–60, 2001.

[106] Brookes, P., rley-Usmar, V. M. Hypothesis: the mitochondrial NO(*) signaling pathway, and the transduction of nitrosative to oxidative cell signals: an alternative function for cytochrome C oxidase. *Free Radic Biol Med* 32(4): pp. 370–4, 2002.

[107] Brookes, P. S., Salinas, E. P., rley-Usmar, K., Eiserich, J. P., Freeman, B. A., rley-Usmar, V. M., Anderson, P. G. Concentration-dependent effects of nitric oxide on mitochondrial permeability transition and cytochrome c release. *J Biol Chem* 275(27): pp. 20474–9, 2000.

[108] Goldenthal, M. J., Marin-Garcia, J. Mitochondrial signaling pathways: a receiver/integrator organelle. *Mol Cell Biochem* 262(1–2): pp. 1–16, 2004.

[109] Jenson, P. K. Antimycin-insensitive oxidation of succinate and reduced nicotinamide adenine dinucleotide in electron transport particles. *Biochem Biophys Acta* (122): pp. 157, 1966.

[110] Hinkle, P. C., Penefsky, H. S., Racker, E. Partial resolution of the enzymes catalyzine oxidative phosphorylation. XII. The H-2-18-O-inorganic phosphate and H-2-18-O-adenosine triphosphate exchange reactions in submitochondrial particles from beef heart. *J Biol Chem* 242(8): pp. 1788–92, 1967.

[111] Grisham, M. B. Reactive metabolites of oxygen and nitrogen in biology and medicine. Austin: R.G.Landes Co, /1.

[112] Brand, M. D., Affourtit, C., Esteves, T. C., Green, K., Lambert, A. J., Miwa, S., Pakay, J. L., Parker, N. Mitochondrial superoxide: production, biological effects, and activation of uncoupling proteins. *Free Radic Biol Med* 37(6): pp. 755–67, 2004.

[113] Manna, S. K., Zhang, H. J., Yan, T., Oberley, L. W., Aggarwal, B. B. Overexpression of manganese superoxide dismutase suppresses tumor necrosis factor-induced apoptosis and activation of nuclear transcription factor-kappaB and activated protein-1. *J Biol Chem* 273(21): pp. 13245–54, 1998.

[114] Alvarez, S., Valdez, L. B., Zaobornyj, T., Boveris, A. Oxygen dependence of mitochondrial nitric oxide synthase activity. *Biochem Biophys Res Commun* 305(3): pp. 771–5, 2003.

[115] Boveris, A., Chance, B. The mitochondrial generation of hydrogen peroxide. General properties and effect of hyperbaric oxygen. *Biochem J* 134(3): pp. 707–16, 1973.

[116] Poderoso, J. J., Carreras, M. C., Lisdero, C., Riobo, N., Schopfer, F., Boveris, A. Nitric oxide inhibits electron transfer and increases superoxide radical production in rat heart mitochondria and submitochondrial particles. *Arch Biochem Biophys* 328(1): pp. 85–92, 1996.

[117] Loschen, G., Flohe, L., Chance, B. Respiratory chain linked H(2)O(2) production in pigeon heart mitochondria. *FEBS Lett* 18(2): pp. 261–4, 1971.

[118] Ceaser, E. K., Ramachandran, A., Levonen, A. L., rley-Usmar, V. M. Oxidized low-density lipoprotein and 15-deoxy-delta 12,14-PGJ2 increase mitochondrial complex I activity in endothelial cells. *Am J Physiol Heart Circ Physiol* 285(6): pp. H2298–308, 2003.

[119] Abadir, P. M., Foster, D. B., Crow, M., Cooke, C. A., Rucker, J. J., Jain, A., Smith, B. J., Burks, T. N., Cohn, R. D., Fedarko, N. S., Carey, R. M., O'Rourke, B., Walston, J. D. Identification and characterization of a functional mitochondrial angiotensin system. *Proc Natl Acad Sci U S A* 108(36): pp. 14849–54, 2011.

[120] Pacher, P., Sharma, K., Csordás, G., Zhu, Y., Hajnóczky, G. Uncoupling of ER-mitochondrial calcium communication by transforming growth factor-beta. *Am J Physiol Renal Physiol* 295(5): pp. F1303–12, 2008.

[121] Dai, D. F., Johnson, S. C., Villarin, J. J., Chin, M. T., Nieves-Cintrón, M., Chen, T., Marcinek, D. J., Dorn, G. W. 2nd, Kang, Y. J., Prolla, T. A., Santana, L. F., Rabinovitch, P. S. Mitochondrial oxidative stress mediates angiotensin II-induced cardiac hypertrophy and Galphaq overexpression-induced heart failure. *Circ Res* 108(7): pp. 837–46, 2011.

[122] Sun, L., Xiao, L., Nie, J., Liu, F. Y., Ling, G. H., Zhu, X. J., Tang, W. B., Chen, W. C., Xia, Y. C., Zhan, M., Ma, M. M., Peng, Y. M., Liu, H., Liu, Y. H., Kanwar, Y. S. p66Shc mediates high-glucose and angiotensin II-induced oxidative stress renal tubular injury via mitochondrial-dependent apoptotic pathway. *Am J Physiol Renal Physiol* 299(5): pp. F1014–25, 2010.

[123] Narayanan, D., Xi, Q., Pfeffer, L. M., Jaggar, J. H. Mitochondria control functional CaV1.2 expression in smooth muscle cells of cerebral arteries. *Circ Res* 107(5): pp. 631–41, 2010.

[124] Viel, E. C., Benkirane, K., Javeshghani, D., Touyz, R. M., Schiffrin, E. L. Xanthine oxidase and mitochondria contribute to vascular superoxide anion generation in DOCA-salt hypertensive rats. *Am J Physiol Heart Circ Physiol* 295(1): pp. H281-, 2008.

[125] Mari, M., Caballero, F., Colell, A., Morales, A., Caballeria, J., Fernandez, A., Enrich, C., Fernandez-Checa, J. C., Garcia-Ruiz, C. Mitochondrial free cholesterol loading sensitizes to TNF- and Fas-mediated steatohepatitis. *Cell Metab* 4(3): pp. 185–98, 2006.

[126] Pollock D. M. How does endothelin induce vascular oxidative stress in mineralocorticoid hypertension? *Clin Sci* (Lond). 2006;110(2): pp. 205–6.

[127] Nozoe, M., Hirooka, Y., Koga, Y., Araki, S., Konno, S., Kishi, T., Ide, T., Sunagawa, K. Mitochondria-derived reactive oxygen species mediate sympathoexcitation induced by angiotensin II in the rostral ventrolateral medulla. *J Hypertens* 26(11): pp. 2176–84, 2008.

[128] De Giusti, V. C., Correa, M. V., Villa-Abrille, M. C., Beltrano, C., Yeves, A. M., de Cingolani, G. E., Cingolani, H. E., Aiello, E. A. The positive inotropic effect of endothelin-1 is mediated by mitochondrial reactive oxygen species. *Life Sci* 83(7–8): pp. 264–71, 2008.

[129] Fernandez-Patron, C. Therapeutic potential of the epidermal growth factor receptor transactivation in hypertension: a convergent signaling pathway of vascular tone, oxidative stress, and hypertrophic growth downstream of vasoactive G-protein-coupled receptors? *Can J Physiol Pharmacol* 85(1): pp. 97–104, 2007.

[130] de Cavanagh, E. M., Ferder, L., Toblli, J. E., Piotrkowski, B., Stella, I., Fraga, C. G., Inserra, F. Renal mitochondrial impairment is attenuated by AT1 blockade in experimental type I diabetes. *Am J Physiol Heart Circ Physiol* 294(1): pp. H456–65, 2008.

[131] Zhang, G. X., Lu, X. M., Kimura, S., Nishiyama, A. Role of mitochondria in angiotensin II-induced reactive oxygen species and mitogen-activated protein kinase activation. *Cardiovasc Res* 76(2): pp. 204–12, 2007.

[132] Gunter, T. E., Pfeiffer, D. R. Mechanisms by which mitochondria transport calcium. *Am J Physiol* 258(5 Pt 1): pp. C755–86, 1990.

[133] Zoratti, M., Szabo, I. The mitochondrial permeability transition. *Biochim Biophys Acta* 1241(2): pp. 139–76, 1995.

[134] Kowaltowski, A. J., Castilho, R. F., Vercesi, A. E. Mitochondrial permeability transition and oxidative stress. *FEBS Lett* 495(1–2): pp. 12–5, 2001.

[135] de Cavanagh, E. M., Ferder, M., Inserra, F., Ferder, L. Angiotensin II, mitochondria, cytoskeletal, and extracellular matrix connections: an integrating viewpoint. *Am J Physiol Heart Circ Physiol* 296(3): pp. H550–60, 2009.

[136] Dikalov, S. Cross talk between mitochondria and NADPH oxidases. *Free Radic Biol Med* 2011 1;51(7): pp. 1289–301.

[137] Ungvari, Z., Labinskyy, N., Gupte, S., Chander, P. N., Edwards, J. G., Csiszar, A. Dysregulation of mitochondrial biogenesis in vascular endothelial and smooth muscle cells of aged rats. *Am J Physiol Heart Circ Physiol* 294(5): pp. H2121–8, 2008.

[138] Postnov, Y. V., Orlov, S. N., Budnikov, Y. Y., Doroschuk, A. D., Postnov, A. Y. Mitochondrial energy conversion disturbance with decrease in ATP production as a source of systemic arterial hypertension. *Pathophysiology* 14(3–4): pp. 195–204, 2007.

[139] Zhang, H., Luo, Y., Zhang, W., He, Y., Dai, S., Zhang, R., Huang, Y., Bernatchez, P., Giordano, F. J., Shadel, G., Sessa, W. C., Min, W. Endothelial-specific expression of mitochondrial thioredoxin improves endothelial cell function and reduces atherosclerotic lesions. *Am J Pathol* 170(3): pp. 1108–20, 2007.

[140] Rodriguez-Iturbe, B., Sepassi, L., Quiroz, Y., Ni, Z., Wallace, D. C., Vaziri, N. D. Association of mitochondrial SOD deficiency with salt-sensitive hypertension and accelerated renal senescence. *J Appl Physiol* 102(1): pp. 255–60, 2007.

[141] Zorov, D. B., Juhaszova, M., Sollott, S. J. Mitochondrial ROS-induced ROS release: an update and review. *Biochim Biophys Acta* 1757(5–6): pp. 509–17, 2006.

[142] Ma, Z. A., Zhao, Z., Turk, J. Mitochondrial Dysfunction and β-Cell Failure in Type 2 Diabetes Mellitus. *Exp Diabetes Res* 2012: p. 703538, 2012.

[143] Zhang, A., Jia, Z., Wang, N., Tidwell, T. J., Yang, T. Relative contributions of mitochondria and NADPH oxidase to deoxycorticosterone acetate-salt hypertension in mice. *Kidney Int* 80(1): pp. 51–60, 2011 Jul.

[144]   Liang, M. Hypertension as a mitochondrial and metabolic disease. *Kidney Int* 80(1): pp. 15–6, 2011.

[145]   Marian, A. J. Mitochondrial genetics and human systemic hypertension. *Circ Res* 108(7): p. 784, 2011.

[146]   Doughan, A. K., Harrison, D. G., Dikalov, S. I. Molecular mechanisms of angiotensin II-mediated mitochondrial dysfunction: linking mitochondrial oxidative damage and vascular endothelial dysfunction. *Circ Res* 102(4): pp. 488–96, 2008.

[147]   Zhang A., Jia Z., Wang N., Tidwell T. J., Yang T. Relative contributions of mitochondria and NADPH oxidase to deoxycorticosterone acetate-salt hypertension in mice. *Kidney Int.* 2011;80(1): pp. 51–60

[148]   Touyz, R. M., Yao, G., Viel, E., Amiri, F., Schiffrin, E. L. Angiotensin II and endothelin-1 regulate MAP kinases through different redox-dependent mechanisms in human vascular smooth muscle cells. *J Hypertens* 22(6): pp. 1141–9, 2004.

[149]   Callera, G. E., Tostes, R. C., Yogi, A., Montezano, A. C., Touyz, R. M. Endothelin-1-induced oxidative stress in DOCA-salt hypertension involves NADPH-oxidase-independent mechanisms. *Clin Sci (Lond)* 110(2): pp. 243–53, 2006.

[150]   Chan, S. H., Wu, K. L., Chang, A. Y., Tai, M. H., Chan, J. Y. Oxidative impairment of mitochondrial electron transport chain complexes in rostral ventrolateral medulla contributes to neurogenic hypertension. *Hypertension* 53(2): pp. 217–27, 2009.

[151]   Chen, L., Tian, X., Song, L. Biochemical and biophysical characteristics of mitochondria in the hypertrophic hearts from hypertensive rats. *Chin Med J (Engl)* 108(5): pp. 361–6, 1995.

[152]   Yang, Q., Kim, S. K., Sun, F., Cui, J., Larson, M. G., Vasan, R. S., Levy, D., Schwartz, F. Maternal influence on blood pressure suggests involvement of mitochondrial DNA in the pathogenesis of hypertension: the Framingham Heart Study. *J Hypertens* 25(10): pp. 2067–73, 2007.

[153]   Rachek, L. I., Grishko, V. I., LeDoux, S. P., Wilson, G. L. Role of nitric oxide-induced mtDNA damage in mitochondrial dysfunction and apoptosis. *Free Radic Biol Med* 40(5): pp. 754–62, 2006.

[154]   Geiszt, M. NAD(P)H oxidases: new kids on the block. *Cardiovasc Res* 71: pp. 289–99, 2006.

[155]   Touyz, R. M., Briones, A. M., Sedeek, M., Burger, D., Montezano, A. C. NOX isoforms and reactive oxygen species in vascular health. *Mol Interv* 11(1): pp. 27–35, 2011.

[156]   Bokoch, G. M., Zhao, T. Regulation of the phagocyte NAD(P)H oxidase by Rac GTPase. *Antioxid Redox Signal* 8(9–10): pp. 1533–48, 2006.

[157]   Guzik, T. J., Chen, W., Gongora, M. C., Guzik, B., Lob, H. E., Mangalat, D., Hoch, N., Dikalov, S., Rudzinski, P., Kapelak, B., Sadowski, J., Harrison, D. G. Calcium-dependent

NOX5 nicotinamide adenine dinucleotide phosphate oxidase contributes to vascular oxidative stress in human coronary artery disease. *J Am Coll Cardiol* 52(22): pp. 1803–9, 2008.

[158]   Leto, T. L., Morand, S., Hurt, D., Ueyama, T. Targeting and regulation of reactive oxygen species generation by Nox family NADPH oxidases. *Antioxid Redox Signal* 11(10): p. 260, 2009.

[159]   Petry, A., Weitnauer, M., Görlach, A. Receptor activation of NADPH oxidases. *Antioxid Redox Signal* 13(4): p. 467, 2010.

[160]   Lassegue, B., Clempus, R. E. Vascular NAD(P)H oxidases: specific features, expression, and regulation. *Am J Physiol Regul Integr Comp Physiol* 285: pp. R277–97, 2003.

[161]   Briones, A. M., Tabet, F., Callera, G. E., Montezano, A. C., Yogi, A., He, Y., Quinn, M. T., Salaices, M., Touyz, R. M. Differential regulation of Nox1, Nox2 and Nox4 in vascular smooth muscle cells from WKY and SHR. *J Am Soc Hypertens* 5(3): pp. 137–53, 2011.

[162]   Ago T., Kuroda J., Kamouchi M., Sadoshima J., Kitazono T. Pathophysiological roles of NADPH oxidase/nox family proteins in the vascular system. *-Review and perspective-Circ J.* 2011;75(8): pp. 1791–800.

[163]   Nisimoto, Y., Tsubouchi, R., Diebold, B. A., Qiao, S., Ogawa, H., Ohara, T., Tamura, M. Activation of NAD(P)H oxidase 1 in tumour colon epithelial cells. *Biochem, J* 415(1): pp. 57–65, 2008.

[164]   Gianni, D., Dermardirossian, C., Bokoch, G. M. Direct interaction between Tks proteins and the N-terminal proline-rich region (PRR) of NoxA1 mediates Nox1-dependent ROS generation. *Eur J Cell Biol* 90(2–3): pp. 164–71, 2011.

[165]   Dutta, S., Rittinger, K. Regulation of NOXO1 activity through reversible interactions with p22 and NOXA1. *PLoS One* 5(5): p. e10478, 2010.

[166]   Fernandes, D. C., Manoel, A. H. O., Wosniak, J., Laurindo, F. R. Protein disulfide isomerise overexpression in vascular smooth muscle cells induces spontaneous preemptive NAD(P)H oxidase activation nad Nox1 mRNA expression: effects of nitrosothiol exposure. *Arch Biochem Biophys* 484(2): pp. 197–204, 2009.

[167]   Lee, M. Y., San Martin, A., Mehta, P. K., Dikalova, A. E., Garrido, A. M., Datla, S. R., Lyons, E., Krause, K. H., Banfi, B., Lambeth, J. D., Lassègue, B., Griendling, K. K. Mechanisms of vascular smooth muscle NAD(P)H oxidase 1 (Nox1) contribution to injury-induced neointimal formation. *Arterioscler Thromb Vasc Biol* 29(4): pp. 480–7, 2009.

[168]   Manea, A., Tanase, L. I., Raicu, M., Simionescu, M. Transcriptional regulation of NADPH oxidase isoforms Nox1 and Nox4, by nuclear factor-kappaB in human aortic smooth muscle cells. *Biochem Biophys Res Commun* 396: pp. 901–7, 2010.

[169]   Tabet, F., Schiffrin, E. L., Callera, G. E., He, Y., Yao, G., Ostman, A., Kappert, K., Tonks, N. K., Touyz, R. M. Redox-sensitive signaling by angiotensin II involves oxidative

inactivation and blunted phosphorylation of protein tyrosine phosphatase SHP-2 in vascular smooth muscle cells from SHR. *Circ Res* 103(2): pp. 149–54, 2008.

[170]   Rathore, R., Zheng, Y. M., Niu, C. F., Liu, Q. H., Korde, A., Ho, Y. S., Wang, Y. X. Hypoxia activates NADPH oxidase to increase [ROS]i and [Ca2+]i through the mitochondrial ROS-PKCepsilon signaling axis in pulmonary artery smooth muscle cells. *Free Radic Biol Med* 45(9): pp. 1223–31, 2008.

[171]   Seshiah, P. N., Weber, D. S., Rocic, P., Valppu, L., Taniyama, Y., Griendling, K. K. Angiotensin II stimulation of NAD(P)H oxidase activity: upstream mediators. *Circ Res* 91: pp. 406–13, 2002.

[172]   Dikalova, A. E., Góngora, M. C., Harrison, D. G., Lambeth, J. D., Dikalov, S., Griendling, K. K. Upregulation of Nox1 in vascular smooth muscle leads to impaired endothelium-dependent relaxation via eNOS uncoupling. *Am J Physiol Heart Circ Physiol* 299(3): pp. H673–9, 2010.

[173]   Yogi, A., Mercure, C., Touyz, J., Callera, G. E., Montezano, A. C., Aranha, A. B., Tostes, R. C., Reudelhuber, T., Touyz, R. M. Renal redox-sensitive signaling, but not blood pressure, is attenuated by Nox1 knockout in angiotensin II-dependent chronic hypertension. *Hypertension* 51(2): pp. 500–6, 2008.

[174]   Basset, O., Deffert, C., Foti, M., Bedard, K., Jaquet, V., Ogier-Denis, E., Krause, K. H. NADPH oxidase 1 deficiency alters caveolin phosphorylation and angiotensin II-receptor localization in vascular smooth muscle. *Antioxid Redox Signal* 11(10): pp. 2371–84, 2009.

[175]   Niu, X. L., Madamanchi, N. R., Vendrov, A. E., Tchivilev, I., Rojas, M., Madamanchi, C., Brandes, R. P., Krause, K. H., Humphries, J., Smith, A., Burnand, K. G., Runge, M. S. Nox activator 1: a potential target for modulation of vascular reactive oxygen species in atherosclerotic arteries. *Circulation* 121(4): pp. 549–59, 2010.

[176]   Touyz, R. M., Chen, X., Tabet, F., Yao, G., He, G., Quinn, M. T., Pagano, P. J., Schiffrin, E. L. Expression of a functionally active gp91phox-containing neutrophil-type NAD(P)H oxidase in smooth muscle cells from human resistance arteries: regulation by angiotensin II. *Circ Res* 90(11): pp. 1205–13, 2002.

[177]   Gupte, S. A., Kaminski, P. M., George, S., Kouznestova, L., Olson, S. C., Mathew, R., Hintze, T. H., Wolin, M. S. Peroxide generation by p47phox-Src activation of Nox2 has a key role in protein kinase C-induced arterial smooth muscle contraction. *Am J Physiol Heart Circ Physiol* 296(4): pp. H1048–57, 2009.

[178]   Han, W., Li, H., Villar, V. A., Pascua, A. M., Dajani, M. I., Wang, X., Natarajan, A., Quinn, M. T., Felder, R. A., Jose, P. A., Yu, P. Lipid rafts keep NADPH oxidase in the inactive state in human renal proximal tubule cells. *Hypertension* 51(2): pp. 481–5, 2008.

[179] Touyz, R. M., Mercure, C., He, Y., Javeshghani, D., Yao, G., Callera, G. E., Yogi, A., Lochard, N., Reudelhuber, T. L. Angiotensin II-dependent chronic hypertension and cardiac hypertrophy are unaffected by gp91phox-containing NAD(P)H oxidase. *Hypertension* 45(4): pp. 530–7, 2005.

[180] Chen, H., Song, Y. S., Chan, P. H. Inhibition of NADPH oxidase is neuroprotective after ischemia-reperfusion. *J Cereb Blood Flow Metab* 29(7): pp. 1262–72, 2009.

[181] Violi, F., Sanguigni, V., Carnevale, R., Plebani, A., Rossi, P., Finocchi, A., Pignata, C., De Mattia, D., Martire, B., Pietrogrande, M. C., Martino, S., Gambineri, E., Soresina, A. R., Pignatelli, P., Martino, F., Basili, S., Loffredo, L. Hereditary deficiency of gp91(phox) is associated with enhanced arterial dilatation: results of a multicenter study. *Circulation* 120(16): pp. 1616–22, 2009.

[182] Loukogeorgakis, S. P., van den Berg, M. J., Sofat, R., Nitsch, D., Charakida, M., Haiyee, B., de Groot, E., MacAllister, R. J., Kuijpers, T. W., Deanfield, J. E. Role of NADPH oxidase in endothelial ischemia/reperfusion injury in humans. *Circulation* 121(21): pp. 2310–6, 2010.

[183] Geiszt, M., Kopp, J. B., Várnai, P., Leto, T. L. Identification of renox, an NAD(P)H oxidase in kidney. *Proc Natl Acad Sci U S A* 97(14): pp. 8010–4, 2000.

[184] Sedeek, M., Callera, G., Montezano, A., Gutsol, A., Heitz, F., Szyndralewiez, C., Page, P., Kennedy, C. R., Burns, K. D., Touyz, R. M., Hébert, R. L. Critical role of Nox4-based NADPH oxidase in glucose-induced oxidative stress in the kidney: implications in type 2 diabetic nephropathy. *Am J Physiol Renal Physiol* 299(6): pp. F1348–58, 2010.

[185] Hilenski, L. L., Clempus, R. E., Quinn, M. T., Lambeth, J. D., Griendling, K. K. Distinct subcellular localizations of Nox1 and Nox4 in vascular smooth muscle cells. *Arterioscler Thromb Vasc Biol* 24(4): pp. 677–83, 2004.

[186] Chen, K., Kirber, M. T., Xiao, H., Yang, Y., Keaney, J. F. Jr. Regulation of ROS signal transduction by NAD(P)H oxidase 4 localization. *J Cell Biol* 181(7): pp. 1129–39, 2008.

[187] Wu, R. F., Ma, Z., Liu, Z., Terada, L. S. Nox4-derived H2O2 mediates endoplasmic reticulum signaling through local Ras activation. *Mol Cell Biol* 30(14): pp. 3553–68, 2010.

[188] Manea, A., Tanase, L. I., Raicu, M., Simionescu, M. Jak/STAT signaling pathway regulates nox1 and nox4-based NADPH oxidase in human aortic smooth muscle cells. *Arterioscler Thromb Vasc Biol* 30(1): pp. 105–12, 2010.

[189] Ismail, A., Sturrock, P., Wu, B., Cahill, K., Norman, T., Huecksteadt, K., Sanders, T. Kennedy and Hoidal, J. NOX4 mediates hypoxia-induced proliferation of human pulmonary artery smooth muscle cells: the role of autocrine production of transforming growth factor-{beta}1 and insulin-like growth factor binding protein-3. *Am J Physiol Lung Cell Mol Physiol* 296: pp. 489–99, 2009.

[190]   Schröder, K., Wandzioch, K., Helmcke, I., Brandes, R. P. Nox4 acts as a switch between differentiation and proliferation in preadipocytes. *Arterioscler Thromb Vasc Biol* 29(2): pp. 239–45, 2009.

[191]   Zhang, M., Brewer, A. C., Schröder, K., Santos, C. X., Grieve, D. J., Wang, M., Anilkumar, N., Yu, B., Dong, X., Walker, S. J., Brandes, R. P., Shah, A. M. NADPH oxidase-4 mediates protection against chronic load-induced stress in mouse hearts by enhancing angiogenesis. *Proc Natl Acad Sci U S A* 107(42): pp. 18121–6, 2010.

[192]   Ray, R., Murdoch, C. E., Wang, M., Santos, C. X., Zhang, M., Alom-Ruiz, S., Anilkumar, N., Ouattara, A., Cave, A. C., Walker, S. J., Grieve, D. J., Charles, R. L., Eaton, P., Brewer, A. C., Shah, A. M. Endothelial Nox4 NADPH oxidase enhances vasodilatation and reduces blood pressure in vivo. *Arterioscler Thromb Vasc Biol* 31(6): pp. 1368–76, 2011.

[193]   Serrander, L., Jaquet, V., Bedard, K., Plastre, O., Hartley, O., Arnaudeau, S., Demaurex, N., Schlegel, W., Krause, K. H. NOX5 is expressed at the plasma membrane and generates superoxide in response to protein kinase C activation. *Biochimie* 89(9): pp. 1159–67, 2007.

[194]   Fulton, D. J. Nox5 and the regulation of cellular function. *Antioxid Redox Signal* 11(10): pp. 2443–52, 2009.

[195]   Jay, D. B., Papaharalambus, C. A., Seidel-Rogol, B., Dikalova, A. E., Lassègue, B., Griendling, K. K. Nox5 mediates PDGF-induced proliferation in human aortic smooth muscle cells. *Free Radic Biol Med* 45(3): pp. 329–35, 2008.

[196]   Montezano, A. C., Paravicini, T. M., Chignalia, A. Z., Yusuf, H., Almasri, M., He, Y., He, G., Callera, G. E., Krause K-H, Lambeth, D., Touyz, R. M. Nicotinamide adenine dinucleotide phosphate reduced oxidase 5 (Nox5) regulation by Angiotensin II and endothelin-1 is mediated via calcium/calmodulin-dependent pathways in human endothelial cells. *Circ Res* 106(8): pp. 1363–73, 2010.

[197]   Wosniak, J. Jr, Santos, C. X., Kowaltowski, A. J., Laurindo, F. R. Cross-talk between mitochondria and NADPH oxidase: effects of mild mitochondrial dysfunction on angiotensin II-mediated increase in Nox isoform expression and activity in vascular smooth muscle cells. *Antioxid Redox Signal* 11(6):1265–9, 2009.

[198]   Montezano, A. C., Burger, D., Ceravolo, G. S., Yusuf, H., Montero, M., Touyz, R. M. Novel Nox homologues in the vasculature: focusing on Nox4 and Nox5. *Clin Sci (Lond)* 120(4): pp. 131–41, 2011.

[199]   Gongora, M. C., Qin, Z., Laude, K., Kim, H. W., McCann, L., Folz, J. R., Dikalov, S., Fukai, T., Harrison, D. G. Role of extracellular superoxide dismutase in hypertension. *Hypertension* 48(3): pp. 473–81, 2006.

[200]   Daiber, A. Redox signaling (cross-talk) from and to mitochondria involves mitochondrial pores and reactive oxygen species. *Biochim Biophys Acta* 1797(6–7): pp. 897–906, 2010.

[201] Lambeth, J. D. NOX enzymes and the biology of reactive oxygen. *Nat Rev Immunol* 4(3): pp. 181–9, 2004.

[202] Cave, A. C., Brewer, A. C., Panicker, A. N., Ray, R., Grieve, D. J., Walker, S., Shah, A. M. NAD(P)H oxidases in cardiovascular health and disease. *Antiox Redox Sig* 8: pp. 691–727, 2006.

[203] Griendling, K. K. NAD(P)H oxidases: new regulators of old functions. *Antioxid Redox Signal* 8(9–10): pp. 1443–5, 2006.

[204] Sorescu, D., Weiss, D., Lassègue, B., Clempus, R. E., Szöcs, K., Sorescu, G. P., Valppu, L., Quinn, M. T., Lambeth, J. D., Vega, J. D., Taylor, W. R., Griendling, K. K. Superoxide production and expression of nox family proteins in human atherosclerosis. *Circulation* 105(12): pp. 1429–35, 2002.

[205] Selemidis, S., Sobey, C. G., Wingler, K., Schmidt, H. H., Drummond GR NAD(P)H oxidases in the vasculature: molecular features, roles in disease and pharmacological inhibition. *Pharmacol Ther* 120(3): pp. 254–91, 2008.

[206] Zhang, R., Harding, P., Garvin, J. L., Juncos, R., Peterson, E., Juncos, L. A., Liu, R. Is forms and functions of NAD(P)H oxidase at the macula densa. *Hypertension* 53(3): pp. 556–63, 2009.

[207] Li, S., Tabar, S. S., Malec, V., Eul, B. G., Klepetko, W., Weissmann, N., Grimminger, F., Seeger, W., Rose, F., Hänze, J. NOX4 regulates ROS levels under normoxic and hypoxic conditions, triggers proliferation, and inhibits apoptosis in pulmonary artery adventitial fibroblasts. *Antioxid Redox Signal* 10(10): pp. 1687–98, 2008.

[208] Tirone, F., Cox, J. A. NAD(P)H oxidase 5 (NOX5) interacts with and is regulated by calmodulin. *FEBS Lett* 581(6): pp. 1202–8, 2007.

[209] Oakley F. D., Smith R. L., Engelhardt J. F. Lipid rafts and caveolin-1 coordinate interleukin-1beta (IL-1beta)-dependent activation of NFkappaB by controlling endocytosis of Nox2 and IL-1beta receptor 1 from the plasma membrane. *J Biol Chem.* 2009;284(48): pp. 33255–64.

[210] Schulz, E., Münzel, T. NOX5, a new "radical" player in human atherosclerosis? *J Am Coll Cardiol* 52(22): pp. 1810–2, 2008.

[211] Hilenski, L. L., Clempus, R. E., Quinn, M. T., Lambeth, J. D., Griendling, K. K. Distinct subcellular localizations of Nox1 and Nox4 in vascular smooth muscle cells. *Arterioscler Thromb Vasc Biol* 24(4): pp. 677–83, 2004.

[212] Takeya, R., Sumimoto, H. Regulation of novel superoxide-producing NAD(P)H oxidases. *Antioxid Redox Signal* 8(9–10): pp. 1523–32, 2006.

[213] Ambasta, R. K., Kumar, P., Griendling, K. K., Schmidt, H. H., Busse, R., Brandes, R. P. Direct interaction of the novel Nox proteins with p22phox is required for the formation of a functionally active NAD(P)H oxidase. *J Biol Chem* 279(44): pp. 45935–41, 2004.

[214] Kawahara, T., Ritsick, D., Cheng, G., Lambeth, J. D. Point mutations in the proline-rich region of p22phox are dominant inhibitors of Nox1- and Nox2-dependent reactive oxygen generation. *J Biol Chem* 280(36): pp. 31859–69, 2005.

[215] Ambasta, R. K., Schreiber, J. G., Janiszewski, M., Busse, R., Brandes, R. P. Noxa1 is a central component of the smooth muscle NAD(P)H oxidase in mice. *Free Radic Biol Med* 41(2): pp. 193–201, 2006.

[216] Cheng, G., Lambeth, J. D. Alternative mRNA splice forms of NOXO1: differential tissue expression and regulation of Nox1 and Nox3. *Gene* 356: pp. 118–26, 2005.

[217] Ibi, M., Matsuno, K., Shiba, D., Katsuyama, M., Iwata, K., Kakehi, T., Nakagawa, T., Sango, K., Shirai, Y., Yokoyama, T., Kaneko, S., Saito, N., Yabe-Nishimura, C. Reactive oxygen species derived from NOX1/NAD(P)H oxidase enhance inflammatory pain. *J Neurosci* 28(38): pp. 9486–94, 2008.

[218] Peng, Y. J., Yuan, G., Jacono, F. J., Kumar, G. K., Prabhakar, N. R. 5-HT evokes sensory long-term facilitation of rodent carotid body via activation of NAD(P)H oxidase. *J Physiol* 576(Pt 1): pp. 289–95, 2006.

[219] Gao, L., Mann, G. E. Vascular NAD(P)H oxidase activation in diabetes: a double-edged sword in redox signalling. *Cardiovasc Res* 82(1): pp. 9–20, 2009.

[220] Suh, S. W., Shin, B. S., Ma, H., Van Hoecke, M., Brennan, A. M., Yenari, M. A., Swanson, R. A. Glucose and NAD(P)H oxidase drive neuronal superoxide formation in stroke. *Ann Neurol* 64(6): pp. 654–63, 2008.

[221] Nistala, R., Whaley-Connell, A., Sowers, J. R. Redox control of renal function and hypertension. *Antioxid Redox Signal* 10(12): pp. 2047–89, 2008.

[222] Brandes, R. P., Schröder, K. Differential vascular functions of Nox family NAD(P)H oxidases. *Curr Opin Lipidol* 19(5): pp. 513–8, 2008.

[223] El-Benna, J., Dang, P. M., Gougerot-Pocidalo, M. A., Marie, J. C., Braut-Boucher, F. p47phox, the phagocyte NAD(P)H oxidase/NOX2 organizer: structure, phosphorylation and implication in diseases. *Exp Mol Med* 41(4): pp. 217–25, 2009.

[224] Maehara, Y., Miyano, K., Sumimoto, H. Role for the first SH3 domain of p67phox in activation of superoxide-producing NAD(P)H oxidases. *Biochem Biophys Res Commun* 379(2): pp. 589–93, 2009.

[225] Montezano, A. C., Callera, G. E., Yogi, A., He, Y., Tostes, R. C., He, G., Schiffrin, E. L., Touyz, R. M. Aldosterone and angiotensin II synergistically stimulate migration in vascular smooth muscle cells through c-Src-regulated redox-sensitive RhoA pathways. *Arterioscler Thromb Vasc Biol* 28(8): pp. 1511–8, 2008.

[226] Block, K., Eid, A., Griendling, K. K., Lee, D. Y., Wittrant, Y., Gorin, Y. Nox4 NAD(P)H oxidase mediates Src-dependent tyrosine phosphorylation of PDK-1 in response to an-

giotensin II: role in mesangial cell hypertrophy and fibronectin expression. *J Biol Chem* 283(35): pp. 24061–76, 2008.

[227] Ebrahimian T., Touyz R. M. Thioredoxin in vascular biology: role in hypertension. *Antioxid Redox Signal.* 2008;10(6): pp. 1127–36.

[228] Sindhu, R. K., Ehdaie, A., Farmand, F., Dhaliwal, K. K., Nguyen, T., Zhan, C. D., Roberts, C. K., Vaziri, N. D. Expression of catalase and glutathione peroxidase in renal insufficiency. *Biochim Biophys Acta* 1743(1–2): pp. 86–92, 2005.

[229] Sui, H., Wang, W., Wang, P. H., Liu, L. S. Effect of glutathione peroxidase mimic ebselen (PZ51) on endothelium and vascular structure of stroke-prone spontaneously hypertensive rats. *Blood Press* 14(6): pp. 366–72, 2005.

[230] Wassmann, S., Wassmann, K., Nickenig, G. Modulation of oxidant and antioxidant enzyme expression and function in vascular cells. *Hypertension* 44(4): pp. 381–6, 2004.

[231] Tajima, M., Kurashima, Y., Sugiyama, K., Ogura, T., Sakagami, H. The redox state of glutathione regulates the hypoxic induction of HIF-1. *Eur J Pharmacol* 606(1–3): pp. 45–9, 2009.

[232] Wong, C. H., Bozinovski, S., Hertzog, P. J., Hickey, M. J., Crack, P. J. Absence of glutathione peroxidase-1 exacerbates cerebral ischemia-reperfusion injury by reducing post-ischemic microvascular perfusion. *J Neurochem* 107(1): pp. 241–52, 2008.

[233] Chung, S. S., Kim, M., Youn, B. S., Lee, N. S., Park, J. W., Lee, I. K., Lee, Y. S., Kim, J. B., Cho, Y. M., Lee, H. K., Park, K. S. Glutathione peroxidase 3 mediates the antioxidant effect of peroxisome proliferator-activated receptor gamma in human skeletal muscle cells. *Mol Cell Biol* 29(1): pp. 20–30, 2009.

[234] Widder, J. D., Guzik, T. J., Mueller, C. F., Clempus, R. E., Schmidt, H. H., Dikalov, S. I., Griendling, K. K., Jones, D. P., Harrison, D. G. Role of the multidrug resistance protein-1 in hypertension and vascular dysfunction caused by angiotensin II. *Arterioscler Thromb Vasc Biol* 27(4): pp. 762–8, 2007.

[235] Mueller, C. F., Wassmann, K., Widder, J. D., Wassmann, S., Chen, C. H., Keuler, B., Kudin, A., Kunz, W. S., Nickenig, G. Multidrug resistance protein-1 affects oxidative stress, endothelial dysfunction, and atherogenesis via leukotriene C4 export. *Circulation* 117(22): pp. 2912–8, 2008.

[236] Zamocky, M., Furtmüller, P. G., Obinger, C. Evolution of catalases from bacteria to humans. *Antioxid Redox Signal* 10(9): pp. 1527–48, 2008.

[237] Wilcox, C. S., Pearlman, A. Chemistry and antihypertensive effects of tempol and other nitroxides. *Pharmacol Rev* 60(4): pp. 418–69, 2008.

[238] Chrissobolis, S., Didion, S. P., Kinzenbaw, D. A., Schrader, L. I., Dayal, S., Lentz, S. R., Faraci, F. M. Glutathione peroxidase-1 plays a major role in protecting against angiotensin II-induced vascular dysfunction. *Hypertension* 51(4): pp. 872–7, 2008.

[239] Ebrahimian, T., Touyz, R. M. Thioredoxin in vascular biology: role in hypertension. *Antioxid Redox Signal* 10(6): pp. 1127–36, 2008.

[240] Redon, J., Oliva, M. R., Tormos, C., Giner, V., Chaves, J., Iradi, A., et al. Antioxidant activities and oxidative stress byproducts in human hypertension. *Hypertension* 41: pp. 1096–101, 2003.

[241] Welch, W. J., Chabrashvili, T., Solis, G., Chen, Y., Gill, P. S., Aslam, S., Wang, X., Ji, H., Sandberg, K., Jose, P., Wilcox, C. S. Role of extracellular superoxide dismutase in the mouse angiotensin slow pressor response. *Hypertension* 48(5): pp. 934–41, 2006.

[242] Zhou, X. J., Vaziri, N. D., Wang, X. Q., Silva, F. G., Laszik, Z. Nitric oxide synthase expression in hypertension induced by inhibition of glutathione synthase. *J Pharmacol Exp Ther* 300(3): pp. 762–7, 2002.

[243] Collins, A. R., Lyon, C. J., Xia, X., Liu, J. Z., Tangirala, R. K., Yin, F., Boyadjian, R., Bikineyeva, A., Praticò D, Harrison, D. G., Hsueh, W. A. Age-accelerated atherosclerosis correlates with failure to upregulate antioxidant genes. *Circ Res* 104(6): pp. e42–54, 2009.

[244] Abudu, N., Miller, J. J., Attaelmannan, M., Levinson, S. S. Vitamins in human arteriosclerosis with emphasis on vitamin C and vitamin E. *Clin Chim Acta* 339(1–2): pp. 11–25, 2004.

[245] Wojcik, M., Burzynska-Pedziwiatr, I., Wozniak, L. A. A review of natural and synthetic antioxidants important for health and longevity. *Curr Med Chem* 17(28): pp. 3262–88, 2010.

[246] Núñez-Córdoba, J. M., Martínez-González, M. A. Antioxidant vitamins and cardiovascular disease. *Curr Top Med Chem* 11(14): pp. 1861–9, 2011.

[247] Masella, R., Di Benedetto, R., Varì R, Filesi, C., Giovannini, C. Novel mechanisms of natural antioxidant compounds in biological systems: involvement of glutathione and glutathione-related enzymes. *J Nutr Biochem* 16(10): pp. 577–86, 2005.

[248] Ndhlala, A. R., Moyo, M., Van Staden, J. Natural antioxidants: fascinating or mythical biomolecules? *Molecules* 15(10): pp. 6905–30, 2010.

[249] Franchini, M., Targher, G., Lippi, G. Serum bilirubin levels and cardiovascular disease risk: a Janus Bifrons? *Adv Clin Chem* 50: pp. 47–63, 2010.

[250] Lin, J. P., Vitek, L., Schwertner, H. A. Serum bilirubin and genes controlling bilirubin concentrations as biomarkers for cardiovascular disease. *Clin Chem* 56(10): pp. 1535–43, 2010.

[251] Schwertner, H. A., Vítek, L. Gilbert syndrome, UGT1A1*28 allele, and cardiovascular disease risk: possible protective effects and therapeutic applications of bilirubin. *Atherosclerosis* 198(1): pp. 1–11, 2008.

[252] Cadenas, E., Sies, H. Oxidative stress: excited oxygen species and enzyme activity. *Adv Enzyme Regul* 23: pp. 217–37, 1985.

[253] Videla, L. A., Fernández, V. Biochemical aspects of cellular oxidative stress. *Arch Biol Med Exp (Santiago)* 21(1): pp. 85–92, 1988.

[254] Giustarini, D., Dalle-Donne, I., Tsikas, D., Rossi, R. Oxidative stress and human diseases: origin, link, measurement, mechanisms, and biomarkers. *Crit Rev Clin Lab Sci* 46(5–6): pp. 241–81, 2009.

[255] Arbogast, S., Ferreiro, A. Selenoproteins and protection against oxidative stress: seleno-protein N as a novel player at the crossroads of redox signaling and calcium homeostasis. *Antioxid Redox Signal* 12(7): pp. 893–904, 2010.

[256] Halliwell, B. Oxidative stress and neurodegeneration: where are we now? *J Neurochem* 97(6):1634–58257, 2006. Segal, B. H., Veys, P., Malech, H., Cowan, M. J. Chronic granulo-matous disease: lessons from a rare disorder. *Biol Blood Marrow Transplant* 17: pp. S123–31, 2011.

[257] Cohen, R. A., Adachi, T. Nitric-oxide-induced vasodilatation: regulation by physiologic s-glutathiolation and pathologic oxidation of the sarcoplasmic endoplasmic reticulum cal-cium ATPase. *Trends Cardiovasc Med* 16(4): pp. 109–14, 2006.

[258] Münzel, T., Daiber, A., Mülsch, A. Explaining the phenomenon of nitrate tolerance. *Circ Res* 97(7): pp. 618–28, 2005.

[259] Mazzanti, L., Raffaelli, F., Vignini, A., Nanetti, L., Vitali, P., Boscarato, V., Giannubilo, S. R., Tranquilli, A. L. Nitric oxide and peroxynitrite platelet levels in gestational hypertension and preeclampsia. *Platelets* 23(1): pp. 26–35, 2012.

[260] Chavez, A., Miranda, L. F., Pichiule, P., Chavez, J. C. Mitochondria and hypoxia-induced gene expression mediated by hypoxia-inducible factors. *Ann N Y Acad Sci* 1147: pp. 312–20, 2008.

[261] Feissner, R. F., Skalska, J., Gaum, W. E., Sheu, S. S. Crosstalk signaling between mitochon-drial Ca2+ and ROS. *Front Biosci* 14: pp. 1197–218, 2009.

[262] Bashan, N., Kovsan, J., Kachko, I., Ovadia, H., Rudich, A. Positive and negative regulation of insulin signaling by reactive oxygen and nitrogen species. *Physiol Rev* 89(1): pp. 27–71, 2009.

[263] Monteiro, H. P., Arai, R. J., Travassos, L. R. Protein tyrosine phosphorylation and protein tyrosine nitration in redox signaling. *Antioxid Redox Signal* 10(5): pp. 843–89, 2008.

[264] Choi H. K., Kim T. H., Jhon G. J., Lee S. Y. Reactive oxygen species regulate M-CSF-induced monocyte/macrophage proliferation through SHP1 oxidation. *Cell Signal.* 2011;23(10): pp. 1633–9.

[265] Callera, G. E., Montezano, A. C., Yogi, A., Tostes, R. C., Touyz, R. M. Vascular signaling through cholesterol-rich domains: implications in hypertension. *Curr Opin Nephrol Hyper-tens* 16(2): pp. 90–104, 2007.

[266] Al Ghouleh, I., Khoo, N. K., Knaus, U. G., Griendling, K. K., Touyz, R. M., Thannickal, V. J., Barchowsky, A., Nauseef, W. M., Kelley, E. E., Bauer, P. M., Darley-Usmar, V., Shiva, S., Cifuentes-Pagano, E., Freeman, B. A., Gladwin, M. T., Pagano, P. J. Oxidases and

peroxidases in cardiovascular and lung disease: new concepts in reactive oxygen species signaling. *Free Radic Biol Med* 51(7): pp. 1271–88, 2011 Oct 1.

[267]   Touyz, R. M. Molecular and cellular mechanisms in vascular injury in hypertension: role of angiotensin II. *Curr Opin Nephrol Hypertens* 14(2): pp. 125–31, 2005.

[268]   Wilcox, C. S. Oxidative stress and nitric oxide deficiency in the kidney: a critical link to hypertension? *Am J Physiol Regul Integr Comp Physiol* 289(4):R913–35, 2005 Oct.

[269]   Hirooka, Y. Role of reactive oxygen species in brainstem in neural mechanisms of hypertension. *Auton Neurosci* 142(1–2): pp. 20–4, 2008.

[270]   Mori, T., Cowley, A. W. Jr, Ito, S. Molecular mechanisms and therapeutic strategies of chronic renal injury: physiological role of angiotensin II-induced oxidative stress in renal medulla. *J Pharmacol Sci* 100(1): pp. 2–8, 2006.

[271]   Mori T., Ogawa S., Cowely A. W. Jr, Ito S. Role of renal medullary oxidative and/or carbonyl stress in salt-sensitive hypertension and diabetes. *Clin Exp Pharmacol Physiol.* 2012;39(1): pp. 125–31.

[272]   Hayden, M. R., Whaley-Connell, A., Sowers, J. R. Renal redox stress and remodeling in metabolic syndrome, type 2 diabetes mellitus, and diabetic nephropathy: paying homage to the podocyte. *Am J Nephrol* 25(6): pp. 553–69, 2005.

[273]   Manning, R. D. Jr, Tian, N., Meng, S. Oxidative stress and antioxidant treatment in hypertension and the associated renal damage. *Am J Nephrol* 25(4): pp. 311–7, 2005.

[274]   Hisaki, R., Fujita, H., Saito, F., Kushiro, T. Tempol attenuates the development of hypertensive renal injury in Dahl salt-sensitive rats. *Am J Hypertens* 18(5 Pt 1): pp. 707–13, 2005.

[275]   Manrique C., Lastra G., Gardner M., Sowers J. R. The renin angiotensin aldosterone system in hypertension: roles of insulin resistance and oxidative stress. *Med Clin North Am.* 2009;93(3): pp. 569–82.

[276]   Chabrashvili, T., Tojo, A., Onozato, M. L., Kitiyakara, C., Quinn, M. T., Fujita, T., Welch, W. J., Wilcox, C. S. Expression and cellular localization of classic NADPH oxidase subunits in the spontaneously hypertensive rat kidney. *Hypertension* 39(2): pp. 269–74, 2002.

[277]   Briones, A. M., Tabet, F., Callera, G. E., Montezano, A. C., Yogi, A., He, Y., Quinn, M. T., Salaices, M., Touyz, R. M. Differential regulation of Nox1, Nox2 and Nox4 in vascular smooth muscle cells from WKY and SHR. *J Am Soc Hypertens* 5(3): pp. 137–53, 2011.

[278]   Zucker, I. H. Novel mechanisms of sympathetic regulation in chronic heart failure. *Hypertension* 48(6): pp. 1005–11, 2006.

[279]   Oliveira-Sales, E. B., Colombari DSA, Davisson, R. L., Kasparov, S., Hirata, E. A., Campos, R. R., Paton JFR. Kidney-induced hypertension depends on superoxide signaling in the rostral ventrolateral medulla. *Hypertension* 56, pp. 290–6, 2010.

[280] Hirooka, Y., Sagara, Y., Kishi, T., Sunagawa, K. Oxidative stress and central cardiovascular regulation. Pathogenesis of hypertension and therapeutic aspects. *Circ J* 74(5): pp. 827–35, 2010.

[281] Campese, V. M., Ye, S., Zhong, H., Yanamadala, V., Ye, Z., Chiu, J. Reactive oxygen species stimulate central and peripheral sympathetic nervous system activity. *Am J Physiol Heart Circ Physiol* 287(2): pp. H695–703, 2004.

[282] Sorce, S., Krause, K. H. NOX enzymes in the central nervous system: from signaling to disease. *Antioxid Redox Signal* 11(10): pp. 2481–504, 2009.

[283] Danson, E. J., Li, D., Wang, L., Dawson, T. A., Paterson, D. J. Targeting cardiac sympatho-vagal imbalance using gene transfer of nitric oxide synthase. *J Mol Cell Cardiol* 46(4): pp. 482–9, 2009.

[284] Nuyt, A. M. Mechanisms underlying developmental programming of elevated blood pressure and vascular dysfunction: evidence from human studies and experimental animal models. *Clin Sci (Lond)* 114(1): pp. 1–17, 2008.

[285] Friese, R. S., Mahboubi, P., Mahapatra, N. R., Mahata, S. K., Schork, N. J., Schmid-Schönbein, G. W., O'Connor, D. T. Common genetic mechanisms of blood pressure elevation in two independent rodent models of human essential hypertension. *Am J Hypertens* 18(5 Pt 1): pp. 633–52, 2005.

[286] Török, J. Participation of nitric oxide in different models of experimental hypertension. *Physiol Res* 57(6): pp. 813–25, 2008.

[287] Kirabo, A., Kearns, P. N., Jarajapu, Y. P., Sasser, J. M., Oh, S. P., Grant, M. B., Kasahara, H., Cardounel, A. J., Baylis, C., Wagner, K. U., Sayeski, P. P. Vascular smooth muscle Jak2 mediates angiotensin II-induced hypertension via increased levels of reactive oxygen species. *Cardiovasc Res* 91(1): pp. 171–9, 2011.

[288] Puddu, P., Puddu, G. M., Cravero, E., Rosati, M., Muscari, A. The molecular sources of reactive oxygen species in hypertension. *Blood Press* 17(2): pp. 70–7, 2008.

[289] Roghair, R. D., Segar, J. L., Volk, K. A., Chapleau, M. W., Dallas, L. M., Sorenson, A. R., Scholz, T. D., Lamb, F. S. Vascular nitric oxide and superoxide anion contribute to sex-specific programmed cardiovascular physiology in mice. *Am J Physiol Regul Integr Comp Physiol* 296(3): pp. R651–62, 2009.

[290] Fukai, T., Ishizaka, N., Rajagopalan, S., Laursen, J. B., Capers, Q. T., Taylor, WRl. p22phox mRNA expression and NAD(P)H oxidase activity are increased in aortas from hypertensive rats. *Circ Res* 80: pp. 45–51, 1997.

[291] Tornavaca, O., Pascual, G., Barreiro, M. L., Grande, M. T., Carretero, A., Riera, M., Garcia-Arumi, E., Bardaji, B., González-Núñez, M., Montero, M. A., López-Novoa, J. M.,

Meseguer, A. Kidney androgen-regulated protein transgenic mice show hypertension and renal alterations mediated by oxidative stress. *Circulation* 119(14): pp. 1908–17, 2009.

[292] Hopps, E., Lo Presti, R., Caimi, G. Pathophysiology of polymorphonuclear leukocyte in arterial hypertension. *Clin Hemorheol Microcirc* 41(3): pp. 209–18, 2009.

[293] Cui W., Matsuno K., Iwata K., Ibi M., Katsuyama M., Kakehi T., Sasaki M., Ikami K., Zhu K., Yabe-Nishimura C. NADPH oxidase isoforms and anti-hypertensive effects of atorvastatin demonstrated in two animal models. *J Pharmacol Sci.* 2009;111(3): pp. 260–8.

[294] Carlström, M., Persson, A. E. Important role of NAD(P)H oxidase 2 in the regulation of the tubuloglomerular feedback. *Hypertension* 53(3): pp. 456–7, 2009.

[295] Haque, M. Z., Majid, D. S. High salt intake delayed angiotensin II-induced hypertension in mice with a genetic variant of NADPH oxidase. *Am J Hypertens* 24(1): p. 114, 2011.

[296] Park, Y. M., Lim, B. H., Touyz, R. M., Park, J. B. Expression of NAD(P)H oxidase subunits and their contribution to cardiovascular damage in aldosterone/salt-induced hypertensive rat. *J Korean Med Sci* 23(6): pp. 1039–45, 2008.

[297] Inaba, S., Iwai, M., Furuno, M., Tomono, Y., Kanno, H., Senba, I., Okayama, H., Mogi, M., Higaki, J., Horiuchi, M. Continuous activation of renin–angiotensin system impairs cognitive function in renin/angiotensinogen transgenic mice. *Hypertension* 53(2): pp. 356–62, 2009.

[298] Haque, M. Z., Majid, D. S. Reduced renal responses to nitric oxide synthase inhibition in mice lacking the gene for gp91phox subunit of NAD(P)H oxidase. *Am J Physiol Renal Physiol* 295(3): pp. F758–64, 2008.

[299] Byrne, J. A., Grieve, D. J., Bendall, J. K., Li, J. M., Gove, C., Lambeth, J. D., Cave, A. C., Shah, A. M. Contrasting roles of NAD(P)H oxidase isoforms in pressure-overload versus angiotensin II-induced cardiac hypertrophy. *Circ Res* 93(9): pp. 802–5, 2003.

[300] Modlinger, P., Chabrashvili, T., Gill, P. S., Mendonca, M., Harrison, D. G., Griendling, K. K., Li, M., Raggio, J., Wellstein, A., Chen, Y., Welch, W. J., Wilcox, C. S. RNA silencing in vivo reveals role of p22phox in rat angiotensin slow pressor response. *Hypertension* 47(2): pp. 238–44, 2006.

[301] Laude, K., Cai, H., Fink, B., Hoch, N., Weber, D. S., McCann, L., Kojda, G., Fukai, T., Schmidt, H. H., Dikalov, S., Ramasamy, S., Gamez, G., Griendling, K. K., Harrison, D. G. Hemodynamic and biochemical adaptations to vascular smooth muscle overexpression of p22phox in mice. *Am J Physiol Heart Circ Physiol* 288(1): pp. H7–12, 2005.

[302] Virdis, A., Neves, M. F., Amiri, F., Touyz, R. M., Schiffrin, E. L. Role of NAD(P)H oxidase on vascular alterations in angiotensin II-infused mice. *J Hypertens* 22: pp. 535–42, 2004.

[303] Hu, L., Zhang, Y., Lim, P. S., Miao, Y., Tan, C., McKenzie, K. U., Schyvens, C. G., Whit-worth, J. A. Apocynin but not L-arginine prevents and reverses dexamethasone-induced hypertension in the rat. *Am J Hypertens* 19(4): pp. 413–8, 2006.

[304] Rey, F. E., Cifuentes, M. E., Kiarash, A., Quinn, M. T., Pagano, P. J. Novel competitive inhibitor of NAD(P)H oxidase assembly attenuates vascular O(2)(-) and systolic blood pressure in mice. *Circ Res* 89: pp. 408–14, 2001.

[305] Gavazzi, G., Banfi, B., Deffert, C., Fiette, L., Schappi, M., Herrmann, F., Krause, K. H. Decreased blood pressure in NOX1-deficient mice. *FEBS Lett* 580(2): pp. 497–504, 2006.

[306] Matsuno, K., Yamada, H., Iwata, K., Jin, D., Katsuyama, M., Matsuki, M., Takai, S., Yamanishi, K., Miyazaki, M., Matsubara, H., Yabe-Nishimura, C. Nox1 is involved in angiotensin II-mediated hypertension: a study in Nox1-deficient mice. *Circulation* 112(17): pp. 2677–85, 2005.

[307] Dikalova, A., Clempus, R., Lassegue, B., Cheng, G., McCoy, J., Dikalov, S. Nox1 over-expression potentiates angiotensin II-induced hypertension and vascular smooth muscle hypertrophy in transgenic mice. *Circulation* 112: pp. 2668–76, 2005.

[308] Bhatia K., Elmarakby A. A., El-Remessey A., Sullivan J. C. Oxidative stress contributes to sex differences in angiotensin II-mediated hypertension in spontaneously hypertensive rats. *Am J Physiol Regul Integr Comp Physiol.* 2012;302(2): pp. R274–82.

[309] Zhang X. H., Lei H., Liu A. J., Zou Y. X., Shen F. M., Su D. F. Increased oxidative stress is responsible for severer cerebral infarction in stroke-prone spontaneously hypertensive rats. *CNS Neurosci Ther.* 2011;17(6): pp. 590–8.

[310] Peixoto, E. B., Pessoa, B. S., Biswas, S. K., Lopes de Faria, J. B. Antioxidant SOD mimetic prevents NAD(P)H oxidase-induced oxidative stress and renal damage in the early stage of experimental diabetes and hypertension. *Am J Nephrol* 29(4):3 pp. 09–18, 2009.

[311] García-Redondo, A. B., Briones, A. M., Beltrán, A. E., Alonso, M. J., Simonsen, U., Salaices, M. Hypertension increases contractile responses to hydrogen peroxide in resistance arteries through increased thromboxane A2, Ca2+, and superoxide anion levels. *J Pharmacol Exp Ther* 328(1): pp. 19–27, 2009.

[312] Takaki, A., Morikawa, K., Murayama, Y., Yamagishi, H., Hosoya, M., Ohashi, J., Shi-mokawa, H. Roles of endothelial oxidases in endothelium-derived hyperpolarizing factor responses in mice. *J Cardiovasc Pharmacol* 52(6): pp. 510–7, 2008.

[313] Yamamoto, E., Tamamaki, N., Nakamura, T., Kataoka, K., Tokutomi, Y., Dong, Y. F., Fu-kuda, M., Matsuba, S., Ogawa, H., Kim-Mitsuyama, S. Excess salt causes cerebral neuronal apoptosis and inflammation in stroke-prone hypertensive rats through angiotensin II-induced NAD(P)H oxidase activation. *Stroke* 39(11): pp. 3049–56, 2008.

[314] Somers, M. J., Mavromatis, K., Galis, Z. S., Harrison, D. G. Vascular superoxide production and vasomotor function in hypertension induced by deoxycorticosterone acetate-salt. *Circulation* 101: pp. 1722–8, 2000.

[315] Callera, G. E., Touyz, R. M., Teixeira, S. A., Muscara, M. N., Carvalho, M. H., Fortes, Z. B. ETA receptor blockade decreases vascular superoxide generation in DOCA-salt hypertension. *Hypertension* 42: pp. 811–7, 2003.

[316] Park, J. B., Touyz, R. M., Chen, X., Schiffrin, E. L. Chronic treatment with a superoxide dismutase mimetic prevents vascular remodeling and progression of hypertension in salt-loaded stroke-prone spontaneously hypertensive rats. *Am J Hypertens* 15(1 Pt 1): pp. 78–84, 2002.

[317] Jiménez, R., López-Sepúlveda, R., Kadmiri, M., Romero, M., Vera, R., Sánchez, M., Vargas, F., O'Valle, F., Zarzuelo, A., Dueñas, M., Santos-Buelga, C., Duarte, J. Polyphenols restore endothelial function in DOCA-salt hypertension: role of endothelin-1 and NADPH oxidase. *Free Radic Biol Med* 43(3): pp. 462–73, 2007.

[318] Elmarakby, A. A., Loomis, E. D., Pollock, J. S., Pollock, D. M. NAD(P)H oxidase inhibition attenuates oxidative stress but not hypertension produced by chronic ET-1. *Hypertension* 45: pp. 283–7, 2005.

[319] Amiri, F., Virdis, A., Neves, M. F., Iglarz, M., Seidah, N. G., Touyz, R. M. Endothelium-restricted overexpression of human endothelin-1 causes vascular remodeling and endothelial dysfunction. *Circulation* 110: pp. 2233–40, 2004.

[320] Quiroz, Y., Ferrebuz, A., Vaziri, N. D., Rodriguez-Iturbe, B. Effect of chronic antioxidant therapy with superoxide dismutase-mimetic drug, tempol, on progression of renal disease in rats with renal mass reduction. *Nephron Exp Nephrol* 112(1): pp. e31–42, 2009.

[321] Castro, M. M., Rizzi, E., Rodrigues, G. J., Ceron, C. S., Bendhack, L. M., Gerlach, R. F., Tanus-Santos, J. E. Antioxidant treatment reduces matrix metalloproteinase-2-induced vascular changes in renovascular hypertension. *Free Radic Biol Med* 46(9): pp. 1298–307, 2009.

[322] Chen, X., Touyz, R. M., Park, J. B., Schiffrin, E. L. Antioxidant effects of vitamins C and E are associated with altered activation of vascular NAD(P)H oxidase and superoxide dismutase in stroke-prone SHR. *Hypertension* 38(3 Pt 2): pp. 606–11, 2001.

[323] Fortuno, A., Olivan, S., Beloqui, O., San Jose, G., Moreno, M. U., Diez, J. Association of increased phagocytic NAD(P)H oxidase-dependent superoxide production with diminished nitric oxide generation in essential hypertension. *J Hypertens* 22: pp. 2169–75, 2004.

[324] Higashi, Y., Sasaki, S., Nakagawa, K., Matsuura, H., Oshima, T., Chayama, K. Endothelial function and oxidative stress in renovascular hypertension. *N Engl J Med* 346: pp. 1954–62, 2002.

[325] Lip, G. Y., Edmunds, E., Nuttall, S. L., Landray, M. J., Blann, A. D., Beevers, D. G. Oxidative stress in malignant and non-malignant phase hypertension. *J Hum Hypertens* 16: pp. 333–6, 2002.

[326] Lee, V. M., Quinn, P. A., Jennings, S. C., Ng, L. L. Neutrophil activation and production of reactive oxygen species in pre-eclampsia. *J Hypertens* 21: pp. 395–402, 2003.

[327] Ward, N. C., Hodgson, J. M., Puddey, I. B., Mori, T. A., Beilin, L. J., Croft, K. D. Oxidative stress in human hypertension: association with antihypertensive treatment, gender, nutrition, and lifestyle. *Free Radic Biol Med* 36: pp. 226–232, 2004.

[328] Ide, T., Tsutsui, H., Ohashi, N., Hayashidani, S., Suematsu, N., Tsuchihashi, M., Tamai, H., Takeshita, A. Greater oxidative stress in healthy young men compared with premenopausal women. *Arterioscler Thromb Vasc Biol* 22(3): pp. 438–42, 2002.

[329] Minuz, P., Patrignani, P., Gaino, S., Seta, F., Capone, M. L., Tacconelli, S., Degan, M., Faccini, G., Fornasiero, A., Talamini, G., Tommasoli, R., Arosio, E., Santonastaso, C. L., Lechi, A., Patrono, C. Determinants of platelet activation in human essential hypertension. *Hypertension* 43: pp. 64–70, 2004.

[330] Yasunari, K., Maeda, K., Nakamura, M., Yoshikawa, J. Oxidative stress in leukocytes is a possible link between blood pressure, blood glucose, and C-reacting protein. *Hypertension* 39: pp. 777–80, 2002.

[331] Lacy, F., Kailasam, M. T., O'Connor, D. T., Schmid-Schonbein, G. W., Parmer, R. J. Plasma hydrogen peroxide production in human essential hypertension: role of heredity, gender, and ethnicity. *Hypertension* 36(5): pp. 878–84, 2000.

[332] Lacy, F., O'Connor, D. T., Schmid-Schönbein, G. W. Plasma hydrogen peroxide production in hypertensives and normotensive subjects at genetic risk of hypertension. *J Hypertens* 16: pp. 291–303, 1998.

[333] Wang, D., Strandgaard, S., Iversen, J., Wilcox, C. S. Asymmetric dimethylarginine, oxidative stress, and vascular nitric oxide synthase in essential hypertension. *Am J Physiol Regul Integr Comp Physiol* 296(2): pp. R195–200, 2009.

[334] Touyz, R. M., Schiffrin, E. L. Increased generation of superoxide by angiotensin II in smooth muscle cells from resistance arteries of hypertensive patients: role of phospholipase D-dependent NAD(P)H oxidase-sensitive pathways. *J Hypertens* 19(7): pp. 1245–54, 2001.

[335] Touyz, R. M., Yao, G., Quinn, M. T., Pagano, P. J., Schiffrin, E. L. p47phox associates with the cytoskeleton through cortactin in human vascular smooth muscle cells: role in NAD(P)H oxidase regulation by angiotensin II. *Arterioscler Thromb Vasc Biol* 25(3): pp. 512–8, 2005.

[336] Zalba, G., San Jose, G., Moreno, M. U., Fortuno, A., Diez, J. NAD(P)H oxidase-mediated oxidative stress: genetic studies of the p22(phox) gene in hypertension. *Antioxid Redox Signal* 7(9–10): pp. 1327–36, 2005.

[337] Moreno, M. U., Jose, G. S., Fortuno, A., Beloqui, O., Diez, J., Zalba, G. The C242T CYBA polymorphism of NAD(P)H oxidase is associated with essential hypertension. *J Hypertens* 24(7): pp. 1299–306, 2006.

[338] Genius, J., Grau, A. J., Lichy, C. The C242T polymorphism of the NAD(P)H oxidase p22phox subunit is associated with an enhanced risk for cerebrovascular disease at a young age. *Cerebrovasc Dis* 26(4): pp. 430–3, 2008.

[339] Kokubo, Y., Iwai, N., Tago, N., Inamoto, N., Okayama, A., Yamawaki, H., Naraba, H., Tomoike, H. Association analysis between hypertension and CYBA, CLCNKB, and KCNMB1 functional polymorphisms in the Japanese population—the Suita Study. *Circ J* 69(2): pp. 138–42, 2005.

[340] Pacher, P., Nivorozhkin, A., Szabó C. Therapeutic effects of xanthine oxidase inhibitors: renaissance half a century after the discovery of allopurinol. *Pharmacol Rev* 58: pp. 87–114, 2006.

[341] Saez, G. T., Tormos, C., Giner, V., Chaves, J., Lozano, J. V., Iradi, A., Redon, J. Factors related to the impact of antihypertensive treatment in antioxidant activities and oxidative stress by-products in human hypertension. *Am J Hypertens* 17(9): pp. 809–16, 2004.

[342] Simic, D. V., Mimic-Oka, J., Pljesa-Ercegovac, M., Savic-Radojevic, A., Opacic, M., Matic, D., Ivanovic, B., Simic, T. Byproducts of oxidative protein damage and antioxidant enzyme activities in plasma of patients with different degrees of essential hypertension. *J Hum Hypertens* 20(2): pp. 149–55, 2006.

[343] Mullan, B. A., Young, I. S., Fee, H., McCance, D. R. Ascorbic acid reduces blood pressure and arterial stiffness in type 2 diabetes. *Hypertension* 40(6): pp. 804–9, 2002.

[344] Zureik, M., Galan, P., Bertrais, S., Mennen, L., Czernichow, S., Blacher, J., Ducimetière, P., Hercberg, S. Effects of long-term daily low-dose supplementation with antioxidant vitamins and minerals on structure and function of large arteries. *Arterioscler Thromb Vasc Biol* 24(8): pp. 1485–91, 2004.

[345] Caner, M., Karter, Y., Uzun, H., Curgunlu, A., Vehid, S., Balci, H., Yucel, R., Güner, I., Kutlu, A., Yaldiran, A., Oztürk, E. Oxidative stress in human sustained and white coat hypertension. *Int J Clin Pract* 60(12): pp. 1565–71, 2006.

[346] Johnstone, E. D., Sawicki, G., Guilbert, L., Winkler-Lowen, B., Cadete, V. J., Morrish, D. W. Differential proteomic analysis of highly purified placental cytotrophoblasts in preeclampsia demonstrates a state of increased oxidative stress and reduced cytotrophoblast antioxidant defense. *Proteomics* 11(20): pp. 4077–84, 2011.

[347] Chen, J., He, J., Hamm, L., Batuman, V., Whelton, P. K. Serum antioxidant vitamins and blood pressure in the United States population. *Hypertension* 40: pp. 810–6, 2002.

[348] Hasnain, B. I., Mooradian, A. D. Recent trials of antioxidant therapy: what should we be telling our patients? *Cleve Clin J Med* 71: pp. 327–34, 2004.

[349]   Jialal, I., Devaraj, S. Antioxidants and atherosclerosis: don't throw out the baby with the bath water. *Circulation* 107: pp. 926–8, 2003.

[350]   Bosch, J., Lonn, E., Pogue, J., Arnold, J. M., Dagenais, G. R., Yusuf, S. Long-term effects of ramipril on cardiovascular events and on diabetes: results of the HOPE study extension. HOPE/HOPE-TOO Study Investigators. *Circulation* 112(9): pp. 1339–46, 2005.

[351]   Schiffrin, E. L. Antioxidants in hypertension and cardiovascular disease. *Mol Interv* 10(6): pp. 354–62, 2010.

[352]   Houston, M. C. Nutraceuticals, vitamins, antioxidants, and minerals in the prevention and treatment of hypertension. *Prog Cardiovasc Dis* 47(6): pp. 396–449, 2005.

[353]   Czernichow, S., Bertrais, S., Blacher, J., Galan, P., Briancon, S., Favier, A., Safar, M., Hercberg, S. Effect of supplementation with antioxidants upon long-term risk of hypertension in the SU.VI.MAX study: association with plasma antioxidant levels. *J Hypertens* 23(11): pp. 2013–8, 2005.

[354]   Myint, P. K., Luben, R. N., Wareham, N. J., Khaw, K. T. Association between plasma vitamin C concentrations and blood pressure in the European prospective investigation into cancer—Norfolk population-based study. *Hypertension* 58(3): pp. 372–9, 2011.

[355]   Bates C. J., Walmsley C. M., Prentice A., Finch S. Does vitamin C reduce blood pressure? Results of a large study of people aged 65 or older. *J Hypertens.* 1998;16(7): pp. 925–932.

[356]   Ward N. C., Hodgson J. M., Croft K. D., Burke V., Beilin L. J., Puddey I. B. The combination of vitamin C and grape seed polyphenols increases blood pressure: a randomized, double-blind, placebo-controlled trial. *J Hypertens* 2005;23(2): pp. 427–434.

[357]   Barbagallo, M., Dominguez, L. J., Tagliamonte, M. R., Resnick, L. M., Paolisso, G. Effects of vitamin E and glutathione on glucose metabolism: role of magnesium. *Hypertension* 34(4 Pt 2): pp. 1002–6, 1999.

[358]   Mishra, G. D., Malik, N. S., Paul, A. A., Wadsworth, M. E., Bolton-Smith, C. Childhood and adult dietary vitamin E intake and cardiovascular risk factors in mid-life in the 1946 British Birth Cohort. *Eur J Clin Nutr* 57(11): pp. 1418–25, 2003.

[359]   Poston, L., Raijmakers, M., Kelly, F. Vitamin E in preeclampsia. *Ann N Y Acad Sci* 1031: pp. 242–8, 2004.

[360]   Rumiris, D., Purwosunu, Y., Wibowo, N., Farina, A., Sekizawa, A. Lower rate of preeclampsia after antioxidant supplementation in pregnant women with low antioxidant status. *Hypertens Pregnancy* 25(3): pp. 241–53, 2006.

[361]   Kelly, R. P., Poo Yeo, K., Isaac, H. B., Lee, C. Y., Huang, S. H., Teng, L., Halliwell, B., Wise, S. D. Lack of effect of acute oral ingestion of vitamin C on oxidative stress, arterial stiffness or blood pressure in healthy subjects. *Free Radic Res* 42(5): pp. 514–22, 2008.

[362] Rumbold, A. R., Crowther, C. A., Haslam, R. R., Dekker, G. A., Robinson, J. S. ACTS Study Group. Vitamins C and E and the risks of preeclampsia and perinatal complications. *N Engl J Med* 354(17): pp. 1796–806, 2006.

[363] Beazley, D., Ahokas, R., Livingston, J., Griggs, M., Sibai, B. M. Vitamin C and E supplementation in women at high risk for preeclampsia: a double-blind, placebo-controlled trial. *Am J Obstet Gynecol* 192(2): pp. 520–1, 2005.

[364] Skyrme-Jones, R. A., O'Brien, R. C., Berry, K. L., Meredith, I. T. Vitamin E supplementation improves endothelial function in type I diabetes mellitus: a randomized, placebo-controlled study. *J Am Coll Cardiol* 36(1): pp. 94–102, 2000.

[365] Cai, H., Griendling, K. K., Harrison, D. G. The vascular NAD(P)H oxidases as therapeutic targets in cardiovascular diseases. *Trends Pharmacol Sci* 24: pp. 471–8, 2003.

[366] Wu, R., Lamontagne, D., de Champlain, J. Antioxidative properties of acetylsalicylic acid on vascular tissues from normotensive and spontaneously hypertensive rats. *Circulation* 105: pp. 387–92, 2002.

[367] Kim, J. A., Neupane, G. P., Lee, E. S., Jeong, B. S., Park, B. C., Thapa, P. NADPH oxidase inhibitors: a patent review. *Expert Opin Ther Pat* 21(8): pp. 1147–58, 2011.

[368] Drummond, G. R., Selemidis, S., Griendling, K. K., Sobey, C. G. Combating oxidative stress in vascular disease: NADPH oxidases as therapeutic targets. *Nat Rev Drug Discov* 10(6): pp. 453–71, 2011.

[369] Dulak, J., Zagorska, A., Wegiel, B., Loboda, A., Jozkowicz, A. New strategies for cardiovascular gene therapy: regulatable pre-emptive expression of pro-angiogenic and antioxidant genes. *Cell Biochem Biophys* 44(1): pp. 31–42, 2006.

[370] Cave, A. Selective targeting of NAD(P)H oxidase for cardiovascular protection. *Curr Opin Pharmacol* 9(2): pp. 208–13, 2009.

[371] Fang, J., Seki, T., Maeda, H. Therapeutic strategies by modulating oxygen stress in cancer and inflammation. *Adv Drug Deliv Rev* 61(4): pp. 290–302, 2009.

[372] Weseler A. R., Bast A. Oxidative stress and vascular function: implications for pharmacologic treatments. *Curr Hypertens Rep.* 2010;12(3): pp. 154–61.

[373] Gupte, S. A. Glucose-6-phosphate dehydrogenase: a novel therapeutic target in cardiovascular diseases. *Curr Opin Investig Drugs* 9(9): pp. 993–1000, 2008.

[374] Huang, H. Y., Caballero, B., Chang, S., Alberg, A. J., Semba, R. D., Schneyer, C. R., Wilson, R. F., Cheng, T. Y., Vassy, J., Prokopowicz, G., Barnes, G. J. 2nd, Bass, E. B. The efficacy and safety of multivitamin and mineral supplement use to prevent cancer and chronic disease in adults: a systematic review for a National Institutes of Health state-of-the-science conference. *Ann Intern Med* 145(5): pp. 372–85, 2006.

[375] Tribble, D. L. Antioxidant consumption and risk of coronary heart disease: emphasis on vitamin C, vitamin E and β-carotene. A statement for the healthcare professionals from the American Heart Association. *Circulation* 99: pp. 591–5, 1999.

[376] Touyz, R. M., Campbell, N., Logan, A., Gledhill, N., Petrella, R., Padwal, R. Canadian Hypertension Education Program. The 2004 Canadian recommendations for the management of hypertension: Part III—Lifestyle modifications to prevent and control hypertension. *Can J Cardiol* 20: pp. 55–83, 2004.

[377] Lopes, H. F., Martin, K. L., Nashar, K., Morrow, J. D., Goodfriend, T. L., Egan, B. M. DASH diet lowers blood pressure and lipid-induced oxidative stress in obesity. *Hypertension* 41(3):422–30, 2003.

[378] John, J. H., Ziebland, S., Yudkin, P., Roe, L. S., Neil, H. A. W. Effects of fruit and vegetable consumption on plasma antioxidant concentrations and blood pressure: a randomized controlled trial. *Lancet* 359: pp. 1969–73, 2002.

[379] Wang, J. S., Lee, T., Chow, S. E. Role of exercise intensities in oxidized low-density lipoprotein-mediated redox status of monocyte in men. *J Appl Physiol* 101(3): pp. 740–4, 2006.

[380] Adams, V., Linke, A., Krankel, N., Erbs, S., Gielen, S., Mobius-Winkler, S., Gummert, J. F., Mohr, F. W., Schuler, G., Hambrecht, R. Impact of regular physical activity on the NAD(P)H oxidase and angiotensin receptor system in patients with coronary artery disease. *Circulation* 111(5): pp. 555–62, 2005.

[381] Pan, Y. X., Gao, L., Wang, W. Z., Zheng, H., Liu, D., Patel, K. P., Zucker, I. H., Wang, W. Exercise training prevents arterial baroreflex dysfunction in rats treated with central angiotensin. *Hypertension* 49(3): pp. 519–27, 2007.

[382] Chen, S., Ge, Y., Si, J., Rifai, A., Dworkin, L. D., Gong, R. Candesartan suppresses chronic renal inflammation by a novel antioxidant action independent of AT1R blockade. *Kidney Int* 74(9): pp. 1128–38, 2008.

[383] Oliveira, P. J., Goncalves, L., Monteiro, P., Providencia, L. A., Moreno, A. J. Are the antioxidant properties of carvedilol important for the protection of cardiac mitochondria? *Curr Vasc Pharmacol* 3(2): pp. 147–58, 2005.

[384] Cifuentes, M. E., Pagano, P. J. Targeting reactive oxygen species in hypertension. *Curr Opin Nephrol Hypertens* 15(2): pp. 179–86, 2006.

[385] Berk, B. C. Novel approaches to treat oxidative stress and cardiovascular diseases. *Trans Am Clin Climatol Assoc* 118: pp. 209–14, 2007.

[386] Sugiura, T., Kondo, T., Kureishi-Bando, Y., Numaguchi, Y., Yoshida, O., Dohi, Y., Kimura, G., Ueda, R., Rabelink, T. J., Murohara, T. Nifedipine improves endothelial function: role of endothelial progenitor cells. *Hypertension* 52(3): pp. 491–8, 2008.

[387]   Dikalov, S., Griendling, K. K., Harrison, D. G. Measurement of reactive oxygen species in cardiovascular studies. *Hypertension* 49(4): pp. 717–27, 2007.

[388]   Tarpey, M. M., Fridovich, I. Methods of detection of vascular reactive species: nitric oxide, superoxide, hydrogen peroxide, and peroxynitrite. *Circ Res* 89(3): pp. 224–36, 2001.

[389]   Ballou, D., Palmer, G., Massey, V. Direct demonstration of superoxide anion production during the oxidation of reduced flavin and of its catalytic decomposition by erythrocuprein. *Biochem Biophys Res Commun* 36(6): pp. 898–904, 1969.

[390]   Li, D. W., Qin, L. X., Li, Y., Nia, R. P., Long, Y. T., Chen, H. Y. CdSe/ZnS quantum dot-Cytochrome c bioconjugates for selective intracellular O2$^{\bullet-}$ sensing. *Chem Commun* 390. Liochev, S. I., Fridovich, I. Lucigenin (bis-N-methylacridinium) as a mediator of superoxide anion production. *Arch Biochem Biophys* 337(1): pp. 115–20, 1997.

[391]   Guzik, T. J., Channon, K. M. Measurement of vascular reactive oxygen species production by chemiluminescence. *Methods Mol Med* 108: pp. 73–89, 2005.

[392]   Tarpey, M. M., White, C. R., Suarez, E., Richardson, G., Radi, R., Freeman, B. A. Chemiluminescent detection of oxidants in vascular tissue. Lucigenin but not coelenterazine enhances superoxide formation. *Circ Res* 84(10): pp. 1203–11, 1999.

[393]   Fink, B., Laude, K., McCann, L., Doughan, A., Harrison, D. G., Dikalov, S. Detection of intracellular superoxide formation in endothelial cells and intact tissues using dihydroethidium and an HPLC-based assay. *Am J Physiol Cell Physiol* 287(4): pp. C895–902, 2004.

[394]   Zhao, H., Joseph, J., Fales, H. M., Sokoloski, E. A., Levine, R. L., Vasquez-Vivar, J., Kalyanaraman, B. Detection and characterization of the product of hydroethidine and intracellular superoxide by HPLC and limitations of fluorescence. *Proc Natl Acad Sci U S A* 102(16): pp. 5727–32, 2005.

[395]   Fernandes, D. C., Wosniak, J. Jr, Pescatore, L. A., Bertoline, M. A., Liberman, M., Laurindo, F. R., Santos, C. X. Analysis of DHE-derived oxidation products by HPLC in the assessment of superoxide production and NADPH oxidase activity in vascular systems. *Am J Physiol Cell Physiol* 292(1): pp. C413–22, 2007.

[396]   Crow, J. P. Dichlorodihydrofluorescein and dihydrorhodamine 123 are sensitive indicators of peroxynitrite in vitro: implications for intracellular measurement of reactive nitrogen and oxygen species. *Nitric Oxide* 1(2): pp. 145–57, 1997.

[397]   Wardman, P. Fluorescent and luminescent probes for measurement of oxidative and nitrosative species in cells and tissues: progress, pitfalls, and prospects. *Free Radic Biol Med* 43(7): pp. 995–1022, 2007.

[398]   Khan, N., Swartz, H. Measurements in vivo of parameters pertinent to ROS/RNS using EPR spectroscopy. *Mol Cell Biochem* 234–235(1–2): pp. 341–57, 2002.

[399] Bardelang, D., Rockenbauer, A., Karoui, H., Finet, J. P., Biskupska, I., Banaszak, K., Tordo, P. Inclusion complexes of EMPO derivatives with 2,6-di-O-methyl-beta-cyclodextrin: synthesis, NMR and EPR investigations for enhanced superoxide detection. *Org Biomol Chem* 4(15): pp. 2874–82, 2006.

[400] Spasojević, I. Free radicals and antioxidants at a glance using EPR spectroscopy. *Crit Rev Clin Lab Sci* 48(3): pp. 114–42, 2011.

[401] Rhodes, C. J. Electron spin resonance. Part one: a diagnostic method in the biomedical sciences. *Sci Prog* 94(Pt 1): pp. 16–96, 2011.

[402] Reginsson, G. W., Schiemann, O. Studying biomolecular complexes with pulsed lectron–electron double resonance spectroscopy. *Biochem Soc Trans* 39(1): pp. 128–39, 2011.

[403] Cai, H., Dikalov, S., Griendling, K. K., Harrison, D. G. Detection of reactive oxygen species and nitric oxide in vascular cells and tissues: comparison of sensitivity and specificity. *Methods Mol Med* 139: pp. 293–311, 2007.

[404] Chiurchiù, V., Maccarrone, M. Chronic inflammatory disorders and their redox control: from molecular mechanisms to therapeutic opportunities. *Antioxid Redox Signal* 15(9): pp. 2605–41, 2011.

Author Biographies

Augusto Montezano, PhD, is a Senior Research Associate in the Kidney Research Centre of the Ottawa Hospital Research Institute (OHRI), University of Ottawa. He obtained his degrees in Pharmacy and his PhD (Pharmacology) at the University of Sao Paulo, Brazil. As a trainee in Brazil he received a highly competitive fellowship from the National Council for Scientific and Technological Development (CNPq) and the Coordenacao de Aperfeicoamento de Pessoal de Nivel Superior (CAPES). In 2005, he joined the OHRI as a fellow and was supported by prestigious fellowships from the Heart and Stroke Foundation of Canada and the Canadian Institutes of Health Research. Dr. Montezano is the recipient of numerous trainee and research awards. He is funded through grants from the Canadian Institutes of Health Research. His major research interest is in the area of free radical biology, signaling and molecular mechanisms of vascular disease. He has published extensively on the topic of Noxs and hypertension and is recognized as an expert in the field.

Rhian M. Touyz, MBBCh, MSc(Med), PhD, is the Director of the Institute of Cardiovascular and Medical Sciences, University of Glasgow and previous Canada Research Chair in Hypertension at the Kidney Research Centre, Ottawa Hospital Research Institute (OHRI)/University of Ottawa. She has received numerous academic and research awards, including the 2012 Robert M. Berne Distinguished Lecturer of the American Physiological Society. Dr Touyz co-chaired the Recommendations Task Force of the Canadian Hypertension Education Program (CHEP), responsible for annual clinical hypertension guidelines. She is on the executive council of the International Society of Hypertension and she is the past President of the Canadian Hypertension Society. She is the current Chair of the Council for High Blood Pressure Research of the American Heart Association. She is funded by grants from the Heart and Stroke Foundation of Canada, the Canadian Institutes of Health Research (NIH equivalent), the Kidney foundation of Canada/Pfizer and the Juvenile Diabetes Research Foundation. She is Editor-in-Chief of *Clinical Science* and Deputy Editor of *Hypertension*. Dr. Touyz has published over 270 original papers and many review articles. Her main focus of research relates to molecular, cellular and vascular mechanisms of hypertension. Her areas of study include clinical and experimental hypertension, signal transduction, oxidative stress, ion transport, vascular biology, adipose tissue biology and diabetes. She has a particular interest in translational research.